MARR

119 Advances in Polymer Science

Thermal and Electrical Conductivity of Polymer Materials

Guest Editors: Y. K. Godovsky and V. P. Privalko

With contributions by
D. M. Bigg, D. Y. Godovsky,
V. V. Novikov, V. P. Privalko,

With 56 Figures and 8 Tables

Springer

Editors:

Prof. Y. K. Godovsky
Karpov Institute of Physical Chemistry
Ul. Obukha 10, 103064 Moscow, RUSSIA

Prof. V. P. Privalko
Academy of Sciences of Ukraine
Institute of Macromolecular Chemistry
253160 Kiev, UKRAINE

ISBN 3-540-58502-8 Springer-Verlag Berlin Heidelberg New York
ISBN 0-387-58502-8 Springer-Verlag New York Berlin Heidelberg

This work is subject to copyright. All rights are reserved, whether the whole or part of the material is concerned, specifically the rights of translation, reprinting, re-use of illustrations, recitation, broadcasting, reproduction on microfilms or in other ways, and storage in data banks. Duplication of this publication or parts thereof is only permitted under the provisions of the German Copyright Law of September 9, 1965, in its current version, and a copyright fee must always be paid.

© Springer-Verlag Berlin Heidelberg 1995
Library of Congress Catalog Card Number 61-642
Printed in Germany

The use of registered names, trademarks, etc. in this publication does not imply, even in the absence of a specific statement, that such names are exempt from the relevant protective laws and regulations and therefore free for general use.

Typesetting: Macmillan India Ltd., Bangalore-25
SPIN: 10470566   02/3020 - 5 4 3 2 1 0 Printed on acid-free paper

Editors

Prof. Akihiro Abe, Department of Industrial Chemistry, Tokyo Institute of Polytechnics, 1583 Iiyama, Atsugi-shi 243-02, Japan
Prof. Henri Benoit, CNRS, Centre de Recherches sur les Macromolécules, 6, rue Boussingault, 67083 Strasbourg Cedex, France
Prof. Hans-Joachim Cantow, Freiburger Materialforschungszentrum, Stefan Meier-Str. 31a, D-79104 Freiburg i. Br., FRG
Prof. Paolo Corradini, Università di Napoli, Dipartimento di Chimica, Via Mezzocannone 4, 80134 Napoli, Italy
Prof. Karel Dušek, Institute of Macromolecular Chemistry, Czech Academy of Sciences, 16206 Prague 616, Czech Republic
Prof. Sam Edwards, University of Cambridge, Department of Physics, Cavendish Laboratory, Madingley Road, Cambridge CB3 OHE, UK
Prof. Hiroshi Fujita, 35 Shimotakedono-cho, Shichiku, Kita-ku, Kyoto 603 Japan
Prof. Gottfried Glöckner, Technische Universität Dresden, Sektion Chemie, Mommsenstr. 13, D-01069 Dresden, FRG
Prof. Dr. Hartwig Höcker, Lehrstuhl für Textilchemie und Makromolekulare Chemie, RWTH Aachen, Veltmanplatz 8, D-52062 Aachen, FRG
Prof. Hans-Heinrich Hörhold, Friedrich-Schiller-Universität Jena, Institut für Organische und Makromolekulare Chemie, Lehrstuhl Organische Polymerchemie, Humboldtstr. 10, D-07743 Jena, FRG
Prof. Hans-Henning Kausch, Laboratoire de Polymères, Ecole Polytechnique Fédérale de Lausanne, MX-D, CH-1015 Lausanne, Switzerland
Prof. Joseph P. Kennedy, Institute of Polymer Science, The University of Akron, Akron, Ohio 44 325, USA
Prof. Jack L. Koenig, Department of Macromolecular Science, Case Western Reserve University, School of Engineering, Cleveland, OH 44106, USA
Prof. Anthony Ledwith, Pilkington Brothers plc. R & D Laboratories, Lathom Ormskirk, Lancashire L40 SUF, UK
Prof. J. E. McGrath, Polymer Materials and Interfaces Laboratory, Virginia Polytechnic and State University Blacksburg, Virginia 24061, USA
Prof. Lucien Monnerie, Ecole Superieure de Physique et de Chimie Industrielles, Laboratoire de Physico-Chimie, Structurale et Macromoléculaire 10, rue Vauquelin, 75231 Paris Cedex 05, France
Prof. Seizo Okamura, No. 24, Minamigoshi-Machi Okazaki, Sakyo-Ku, Kyoto 606, Japan
Prof. Charles G. Overberger, Department of Chemistry, The University of Michigan, Ann Arbor, Michigan 48109, USA
Prof. Helmut Ringsdorf, Institut für Organische Chemie, Johannes-Gutenberg-Universität, J.-J.-Becher Weg 18-20, D-55128 Mainz, FRG
Prof. Takeo Saegusa, KRI International, Inc. Kyoto Research Park 17, Chudoji Minamima-chi, Shimogyo-ku Kyoto 600 Japan
Prof. J. C. Salamone, University of Lowell, Department of Chemistry, College of Pure and Applied Science, One University Avenue, Lowell, MA 01854, USA
Prof. John L. Schrag, University of Wisconsin, Department of Chemistry, 1101 University Avenue. Madison, Wisconsin 53706, USA
Prof. G. Wegner, Max-Planck-Institut für Polymerforschung, Ackermannweg 10, Postfach 3148, D-55128 Mainz, FRG

Preface

Rapid development of modern advanced technologies require new engineering materials with special properties. It is highly improbable that all such requirements can be satisfied by any single-component material; hence, one should look for a multi-component composite with a desired combination of properties. This is a familiar materials science problem which may be solved either by empirical screening of many test specimens, or by theoretical analysis of an appropriate model, or by both. In the latter case, however, a misfit between theory and experiment is not rare; in fact, simple variation of shape and/or size of particles of a disperse component sometimes leads to profound changes in properties of a composite, which the current theoretical models can hardly predict. Thus, the need for the incorporation of new physical ideas gained from experiments into theoretical models becomes obvious.

These considerations were implicit in choosing the contributions for this volume on "High Thermal Conductivity and Low Electrical Conductivity" within the Springer series "Advances in Polymer Science". The paper by D. M. Bigg entitled "Thermal Conductivity of Heterophase Polymer Compositions" gives convincing examples of the effect of particle shape on thermal conductivity of filled polymers, especially at high filler loadings. V. P. Privalko and V. V. Novikov in "Model Treatments of the Heat Conductivity of Heterogeneous Polymers" show how the effect op particle size may be theoretically accounted for within the framework of a realistic model assuming the formation of a matrix "boundary interphase" around disperse particles. The other aspect of particle size issue is addressed by D. Yu. Godovsky in "Electron Behavior and Magnetic Properties of Polymer Nancomposites"; namely "nano-size" particles of certain metals dispersed in al polymer matrix exhibit unusual electron transport, optical, magnetic, and catalytic behavior, thus making the corresponding "nanocomposites" attractive for many practical applications.

It is hoped that the experimental data and corresponding theoretical approaches analyzed in this volume will stimulate further progress in the understanding of electron- and heat transport phenomena in polymer composites

December 1994 Yu. K. Godovsky
V. P. Privalko

Table of Contents

Thermal Conductivity of Heterophase Polymer Compositions
D. M. Bigg 1

Model Treatments of the Heat Conductivity of Heterogeneous Polymers
V. P. Privalko, V. V. Novikov 31

Electron Behavior and Magnetic Properties Polymer-Nanocomposites
D. Y. Godovsky 79

Author Index Volumes 101 - 119 123

Subject Index 129

Thermal Conductivity of Heterophase Polymer Compositions

D.M. Bigg
R.G. Barry Corporation, P.O. Box 129, Columbus, OH 43216, USA

This document is a review of the important aspects of the thermal conductivity of filled polymer compositions. Included in this review is an analysis of the various theories developed to model the behavior of heterogeneous two phase systems. A number of second order models were found to provide good estimates of actual data for composites filled with spherical and irregularly shaped particles. In those cases where the models and data disagreed, it was usually because the packing arrangements in the real systems did not match the assumed packing arrangements in the models. This usually occurred at high filler volume fractions where filler packing depends on mixing, particle size and size distribution, and molding considerations. Similar disparities between real composite systems and models for flake and fiber filled composites made analysis of models for such systems difficult to assess. Experimental methods were reviewed, with the inclusion of the more recently developed unsteady state techniques. Finally, discussion is presented on the use of thermally conductive compositions in heat exchangers where the thermal conductivity is of primary importance to the performance of the device.

List of Symbols and Abbreviations 2

1 **Introduction** 4
2 **Theory** 5
   2.1 Spherical Filler Particles 6
   2.2 Irregularly Shaped Particles 16
   2.3 Flakes 18
   2.4 Short Fibers 19
   2.5 Long Fibers 22
   2.6 Continuous Fibers 22
3 **Measurement Techniques** 23
   3.1 Steady State Techniques 23
   3.2 Unsteady State Approaches 24
4 **Applications** 26
5 **Discussion** 28
6 **References** 29

List of Symbols and Abbreviations

b	parameter in Eq. (18) defined in Eq. (19)
d	constant in Eqs. (5, 8, 9, 11, 12 and 18)
k_c	thermal conductivity of a two phase composite
k_f	thermal conductivity of the filler material
k_m	thermal conductivity of the matrix material
l	fiber length
l	wire strip length
n	power law exponent in Eq. (20)
q	heat flux
q'	rate of heat generated per length of wire in line source technique
r	radial distance from line source
t	time
t_1	time 1
t_2	time 2
x	distance
x_o	material thickness
Δx	change in distance
A	cross-sectional area
A	coefficient in Eq. (15)
A_n	summation parameter in Eq. (6)
A_1	parameter defined in Eq. (7)
A_2	parameter defined in Eq. (8)
A_3	parameter defined in Eq. (9)
B	coefficient in Eq. (15), defined in Eq. (17)
B_n	summation parameter in Eq. (10)
C_1	parameter in Eq. (20) related to effect of filler on crystallinity of polymer
C	parameter in Eq. (20) related to critical structural concentration of filler particles
C_p	specific heat
D	fiber diameter
E	modulus
F	force
H	parameter in Eq. (34), defined in Eq. (35)
J	parameter in Eq. (30), defined in Eq. (31)
L	effective diameter of a flake
N	coefficient in Eq. (13)
Q	rate of heat flow
R	line resistance
S	coefficient in Eqs. (14, 25)
S_{11}	parameter in Eq. (30), defined in Eq. (32)

S_{33}	parameter in Eq. (30), defined in Eq. (33)
T	temperature
T_0	initial sample temperature
T_1	temperature at time t_1
T_2	temperature at time t_2
T_i	constant surface temperature in Eq. (39)
V	average ac line voltage
$V_{3,1}$	average rms voltage at harmonic frequency $3\omega_1$
$V_{3,2}$	average rms voltage at harmonic frequency $3\omega_2$
X	effective thickness of a flake
β	parameter in Eq. (11), defined in Eq. (12)
ε	strain
ϕ_c	critical volume fraction for interparticle network formation
ϕ_f	volume fraction of filler phase
ϕ_m	volume fraction of matrix phase
ϕ_{max}	maximum packing fraction
ρ	density
σ	stress
ψ	parameter in Eq. (15), defined in Eq. (16)
ω	current frequency in radians/sec

1 Introduction

The importance of thermal conductivity in polymers and polymer composites has increased in recent years due to the need for appreciable levels of thermal conductance in circuit boards, heat exchangers, appliances, and machinery. It is also important to know the thermal conductivity of filled compositions in order to model and analyze the heat transfer process during molding. While this is true for the growing range of filled polymers that are molded, it is particularly important in the growing fields of injection molded ceramic and metal powders. As a result of these needs, new developments have been made in devices to measure thermal conductivity, theoretical developments have been revisited, and new data generated to eliminate some of the questions raised regarding the accuracy of older data. According to Mottram there are seven requirements that must be met in order for thermal conductivity data to be considered useful for evaluating theoretical models [1].

1. Detailed information must be presented on the processing of the material and the preparation of the test samples.
2. The data must encompass a wide range of practical filler volume concentrations.
3. Data on the thermal conductivity of the pure components must be available over the temperature range in which the composite is evaluated.
4. A reliable, reproducible experimental technique must be used to measure the thermal conductivity of the test samples.
5. The interface between the matrix and particulate filler must be continuous without voids.
6. The composite should consist of only a matrix and dispersed particulate phase.
7. The size, shape, and spatial distribution of the dispersed phase must be the same as that assumed in any theoretical development.

Despite the recent interest in thermal conductivity, it remains the least studied of the major transport properties, viscosity and mass transfer being much more widely investigated. Transport properties are coefficients that define the ratio between a flux and the directional gradient of the driving force causing heat, momentum, or mass transport. In the case of thermal transport the flux is the heat flux, q, the directional driving force, dT/dx, is the temperature gradient through some distance. q is the rate of heat flow, Q, across a given cross-sectional area, A. Mathematically, the definition of thermal conductivity is:

$$k = q/(dT/dx) = (Q/A)/(\Delta T/x_o). \tag{1}$$

The important consideration that needs to be accounted for is the dependence of thermal conductivity on temperature, pressure, and, in the case of composites, the filler properties of concentration, type, geometry, and orientation. The primary emphasis of this review is to examine the effect of fillers on the thermal conductivity of polymers.

2 Theory

There have been many theoretical models developed describing the thermal conductivity of two phase, particulate filled composites. The number of distinct theoretical equations is in the hundreds, and cannot be reasonably analyzed in any single investigation. Fortunately, periodic reviews have been published to provide critical analysis of the most useful models [2, 3, 4, 5]. In some respects modeling the thermal conductivity of filled systems is easier than modeling the effect of fillers on the other transport properties, viscosity and diffusivity, since the thermal conductivity is measured on a static solid-solid composite, rather than a material system in which one or more of the materials is changing spatially with time. Particle migration is a particularly difficult problem in rheological evaluations of solid-fluid compositions. Various investigators have shown that the dispersion of solid particles in a fluid matrix changes during the imposition of shear, a necessary requirement for measuring viscosity [6, 7]. Moreover, the dispersed phase may develop an internal structure, even at relatively low filler concentrations, a phenomenon which further complicates the rheological analysis. Another complication is that neighboring particles may interact with one another or with the matrix to magnify their relative effect on viscosity, even though a network has not been established. Mass transport is controlled by a variety of phenomena, and is highly nonlinear, making the measurement of diffusivity a complex matter.

The material properties that have been observed to respond most similarly to thermal conductivity over at least a limited range of variables are viscosity, modulus and dielectric strength. Hence, by analogy to these more widely measured properties, models have been adapted to predict the behavior of fillers on the thermal conductivity of polymers. In fact, many of the models developed to predict thermal conductivity have been adapted from models developed to predict one of these other properties. There are few models developed solely to describe the thermal conductivity of composites from thermal transport principles. Care must be taken in adapting models originally developed to describe other properties. Analogies based on the viscosity of filled systems are often inappropriate because solid particulate fillers have an infinite viscosity relative to that of the matrix, while the relative values of thermal conductivity of the two phases are not always so extreme. There is a wide range in possible thermal conductivities of filler particles relative to polymer matrices. Metals have thermal conductivities of the order of 1000 times higher than that of most polymers, while many inorganic fillers have thermal conductivities less than an order of magnitude greater than polymers. Viscosity-based analogies are often valid for metal filled composites, since the effect of relative thermal conductivity between the two phases is minimal once the relative thermal conductivity exceeds 100:1. Also of interest are polymer blends in which the relative values of the thermal conductivities of the two phases are similar, in which case viscosity models based on solid particulate fillers are clearly inappropriate.

The most commonly adapted models have been those developed to describe the modulus of two phase systems. Modulus, E, is not a transport property, but is defined in the same mathematical form as a transport property, and has the added benefit that it also represents a solid-solid composite. Modulus is defined by the following equation:

$$E = \sigma/\varepsilon = (F/A)/(\Delta x/x_0). \tag{2}$$

In this equation σ is the stress (a force, F, imposed over the cross-sectional area A), and ε is the resulting deformation, or strain (increase in dimension, Δx, relative to the original dimension, x_0) of the material. Note the obvious analogy in form between Eqs. (1) and (2). There are three common moduli used to describe materials–tensile, shear, and bulk; these moduli are related to one another through the Poisson's ratio of the material.

The geometry and orientation of filler particles has a significant influence on the thermal conductivity of a composite material. Many of the theoretical treatments are valid for only specific types of filler particles and composite constructions. The possible scenarios include spherical fillers, irregularly shaped fillers, flakes (3-D random and random in-plane), short fibers (3-D random and random in-plane), long fibers (unidirectional and random in-plane), and continuous fibers (unidirectional and cross-ply laminated). Each of these scenerios needs to be addressed separately, and will be done so in the following sections.

2.1 Spherical Filler Particles

There are two basic theoretical approaches that have been followed to model the effect of spherical filler particles on a matrix. These two scenarios are shown schematically in Fig. 1. The first approach allows for the formation of an internal network structure within the matrix. This model considers the contribution of each phase separately, and uses percolation theory to arrive at an effective thermal conductivity for the composite [8–10]. The models which account for particle-particle interactions tend to maximize the effect of the dispersed phase. For example the simplest interactive model is the simple rule-of-mixtures:

$$\frac{k_c}{k_m} = \phi_m + \frac{k_f \phi_f}{k_m} \tag{3}$$

where ϕ is the volume fraction of the dispersed phase, k_m is the thermal conductivity of the matrix, and k_f is the thermal conductivity of the filler. This equation treats each component as contributing to the thermal conductivity of

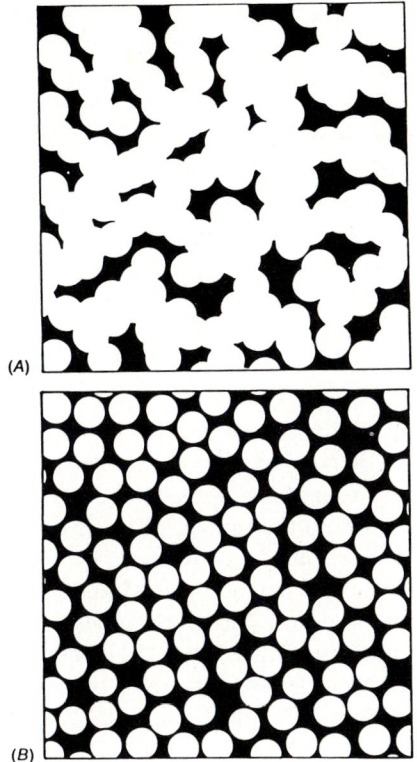

Fig. 1a, b. Basic conceptual models for predicting thermal conductivity of two phase systems: **a** fully interpenetrating dispersed; **b** non-penetrating dispersed phase. Reprinted with permission of VCH Publishers from [5]

the composite in an amount equal to the volume fraction of that component. This clearly maximizes the influence of the minor component. This model is often referred to as a series model, based on the analogy of electrical current flow in a series circuit.

The second approach to modeling the thermal conductivity of a composite is to assume that the composite material responds as a homogeneous material in which each filler particle is an isolated entity. This approach results in a weighted average composite thermal conductivity that inherently includes the effect of the geometry of the filler particles. Most of the theoretical models used to describe or predict the effect of a filler on the thermal conductivity of a homogeneous matrix are based on an averaging technique. Such models assume that there is no connecting network among the filler particles and that each particle exhibits an isolated effect minimizing the effect of the dispersed phase on the thermal conductivity of the matrix, particularly at lower filler concentrations. A prototypical model of this sort follows a parallel electrical circuit arrangement that is represented mathematically as follows:

$$\frac{k_c}{k_m} = \frac{1/k_m}{(\phi_m/k_m + \phi_f/k_f)}. \tag{4}$$

These two models, being simple averages, are referred to as first order models by Mottram. Figure 2 shows how these two simple models predict the thermal conductivity of a composite filled with spherical particles having a ratio of k_f/k_m equal to 1000. Data for such compositions are also included in this Figure. Clearly the data are not described by either model, but do provide upper and lower bounds that encompass the data. The non-interactive (lower bound) model provides a closer match to the data than the rule-of-mixtures (upper bound) model, thus giving impetus to the development of a class of models based on the assumption of isolated filler particles. Efforts such as those of Bruggeman resulted in the following equation [11]:

$$\phi_m \frac{[k_m - k_c]}{k_m + (d-1)k_c} + \phi_f \frac{[k_f - k_c]}{k_f + (d-1)k_c} = 0 \tag{5}$$

where d is a constant that accounts for the dimensionality of the system, 3 in the case of an isotropic material. This equation predicts a percolation threshold at a volume fraction of the dispersed phase at 0.33, a phenomena that has not been observed experimentally except in systems in which the matrix phase is not a continuum. This is despite the fact that the formation of a continuous network of filler particles develops at filler concentrations well below the maximum filler loading. Gurland showed that a network of randomly dispersed spheres developed in a polymer matrix at a volume concentration of 35% [12]. This was demonstrated by a sharp discontinuity in the electrical resistance of the com-

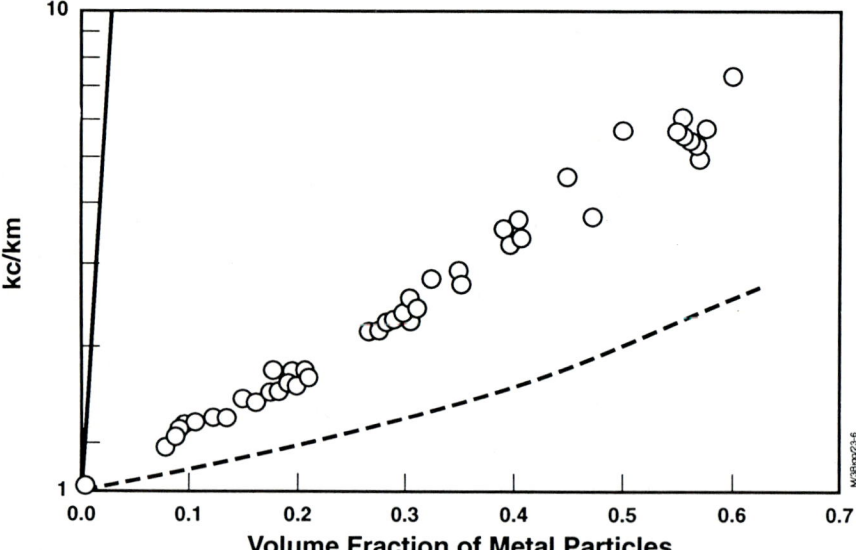

Fig. 2. First order upper and lower bound predictions of the relative thermal conductivity of a two phase system as a function of volume fraction of the dispersed phase in which the particles are spherical and $k_f/k_m \geq 1000$. Data taken from [17, 40–45]

posite when the dispersed phase consisted of metallic spheres. The existence of a critical percolation threshold for electrical conductivity has since been demonstrated for a wide variety of fillers, all at concentrations below the maximum packing fraction [13]. No discontinuity occurs in the thermal conductivity data presented in Fig. 2. This suggests that in most cases the presence of a network of filler particles does not change the basic mechanism of thermal transport in a composite system. According to Torquado, a thermal transport network develops only at the maximum packing fraction. Since the mechanism of thermal transport is far different from that of electrical transport, the definitions of a conductive network are different.

For simplicity in discussing the various models, the model described by Eq. (4) will be referred to as the first order lower bound model, or FOLBM. In order to provide a better fit to existing data a more complex averaging approach is needed. Early attempts to develop such models began with statistically summing the perturbations around each filler particle to calculate the average thermal conductivity of the composition. In general, the relative thermal conductivity of a particle-in-matrix, dispersed two phase system can be expressed as a series expansion of the localized average thermal conductivity determined by localized heat transfer considerations. This expansion is [14]:

$$k_c/k_m = 1 + \sum_{n=1}^{\infty} A_n [k_f - k_m)/k_m]^n \tag{6}$$

where A_n describes the local field, with the three values of A_n given as follows for a system in which the inclusions are spherical particles.

$$A_1 = \phi_f \tag{7}$$

$$A_2 = \phi_m \phi_f / d \tag{8}$$

$$A_3 = \frac{\phi_m \phi_f}{d} [\phi_m + (d-1)\phi_f] \tag{9}$$

where $d = 3$ for an isotropic system. Using only the first three series of this expansion, the equation breaks down when k_f/k_m is greater than 2. Additional coefficients of A_n are needed to treat systems in which there are greater differences in the thermal conductivity of the two phases. This cannot be readily accomplished since greater knowledge is required about the localized microstructure than is available. Most real systems involve fillers and matrices having thermal conductivity ratios greater than 2 and therefore Eq. (6) is of little practical use.

Another series expansion approach that accounts for a wide range of k_m and k_f has the following form [15]:

$$k_c/k_m = 1 + \sum_{n=1}^{\infty} B_n \phi_f^n. \tag{10}$$

In this equation B_n is a multidimensional integral to the boundary value heat conduction problem around the particle n, weighted according to the size distribution of the particles dispersed in the matrix. Various values of B_n have been determined for several idealized fillers. Unfortunately, this approach breaks down at high values of filler concentration.

The main problem with the statistical approaches is that more information is needed about the microstructure of the systems than is usually available. The scatter common in thermal conductivity data at higher filler loadings attests to this lack of knowledge about the localized microstructure. One way around this dilemma is to develop a system of equations which provides upper and lower bounds on the possible responses that can be expected without having such detailed knowledge of the dispersed phase microstructure. Several second order models have been developed based on this approach.

Hashin and Shtrikman produced the following lower bound equation to describe the effect of spherical filler particles on the thermal conductivity of a randomly dispersed, particle-in-matrix, two phase system [16]:

$$\frac{k_c}{k_m} = \frac{1 + (d-1)\phi\beta}{1 - \phi\beta} \tag{11}$$

where

$$\beta = \frac{k_f - k_m}{k_f + (d-1)k_m} \tag{12}$$

and $d = 3$ for spherical filler particles.

Hamilton and Crosser produced the following lower bound equation [17]:

$$\frac{k_c}{k_m} = \frac{k_f + Nk_m - N\phi(k_m - k_f)}{k_f + Nk_m + \phi(k_m - k_f)} \tag{13}$$

in which N is an empirical constant that is equal to 2 for spherical filler particles. Hatta and Taya developed a model based on the analogy of thermal conductivity to modulus that was originally developed by Eshelby [18, 19]. The equation Hatta and Taya arrived at is:

$$\frac{k_c}{k_m} = 1 + \frac{\phi}{S(1-\phi) + k_m/(k_f - k_m)} \tag{14}$$

where $S = 1/3$ for spherical filler particles. After algebraic manipulation, these two equations, Eqs. (11), (13) and (14), can be shown to be equivalent for spherical filler particles. These equations are weighted averages of the thermal conductivities and volume concentrations of the two components, but in a more complex averaging scheme than the simple first order model presented in Eq. (4). They are second order models in that they can be derived from Eq. (6) through to the second coefficient A_n. These models will be referred to as SOLBM.

Nielsen developed a model for thermal conductivity, based on the Kerner equations describing the modulus of two phase systems, that provides a different response that of the SOLBM [20]. Nielsen's model added to the consideration of the relative conductivity of the two phases the effect of the maximum packing fraction of the filler, ϕ_{max}. Moreover, the model readily adjusts to handle non-spherical particles through a coefficient that is dependent on the particle shape and orientation. The following equations were part of this model, which was found to be quite useful by a number of investigators [2, 3]:

$$\frac{k_c}{k_m} = \frac{1 + (A - 1)B\phi}{(1 - \psi B\phi)} \tag{15}$$

$$\psi = 1 + \frac{(1 - \phi_{max})\phi}{\phi_{max}^2} \tag{16}$$

$$B = \frac{k_f/k_m - 1}{k_f/k_m + A - 1}. \tag{17}$$

The coefficient A depends on the geometry and orientation of the filler particles. Nielsen provided values of A for a wide range of common filler types, many of which are listed in Table 1. For spherical filler particles, A = 2.5. Nielsen's model has a form similar to that of Hashin and Shtrikman. The major modification is the addition of a term to account for the maximum packing fraction. This model has sufficient flexibility and basis in materials properties to describe a wide range of actual data. Including the maximum packing fraction in the model has the effect of amplifying concentration effects at high filler loadings. This is consistent with observations that predictions become difficult to make as the filler concentration approaches the maximum packing fraction. Recall that Torquado discussed the development of a thermal transport network at the maximum packing fraction. The development of such a network is not always sharply defined. It depends on the particle size distribution and microscale packing developed in the composite. Figure 3 shows the predicted effect of maximum packing fraction on the relative thermal conductivity of composites filled with spherical metal particles ($k_f/k_m \geq 1000$). Different maximum packing fractions are possible with spherical fillers due to differences in either packing arrangement or particle size distribution. A value of 0.637 has been derived for the maximum packing of monodisperse particles randomly dispersed in a

Table 1. Values of "A" for common filler types

Filler	1/D	A
spheres	1	2.5
isotropic particles	1	3.5
random fibers, 3D	2	2.58
random fibers, 3D	4	3.08
random fibers, 3D	6	3.88
random fibers, 3D	10	5.93
random fibers, 3D	15	9.83
random fibers, 3D	35	30.00
uniaxially oriented fibers, parallel	–	2L/D
uniaxially oriented fibers, perpendicular	–	0.5

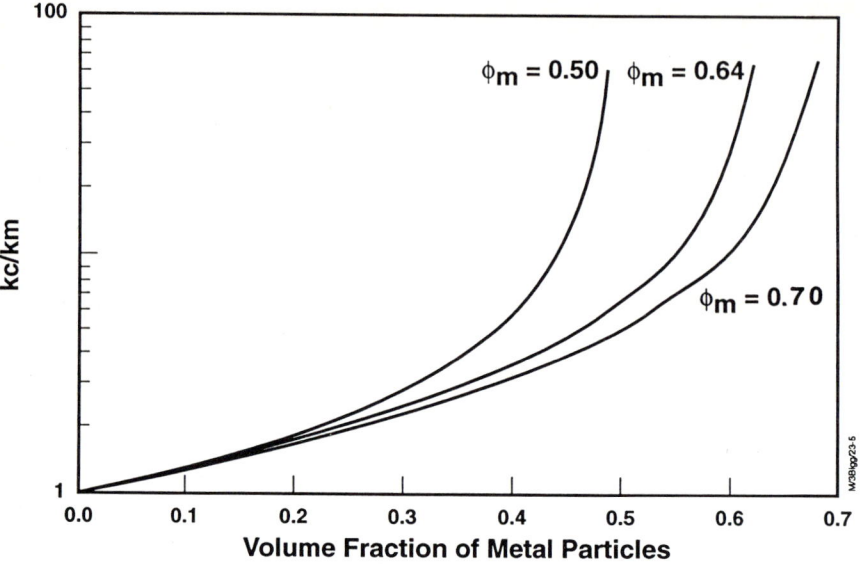

Fig. 3. The predicted effect of maximum packing fraction on the relative thermal conductivity of composites filled with spherical particles in which $k_f/k_m \geq 1000$, according to the Nielsen model

matrix. A broad, or bimodal, particle size distribution will result in a higher value. Table 2 contains a list of the maximum packing fraction for a number of filler materials and packing arrangements.

Figure 4 is a comparison of the predictions of the SOLBM models of Hatta and Taya, Hamilton and Crosser, Hashin and Shtrikman with that of Nielsen for composites filled with spherical metal particles in which the relative thermal conductivity of the dispersed phase is approximately 1000 times greater than that of the matrix. All of these models are in excellent agreement up to a volume fraction of 0.30; the variation between Nielsen's model and the other SOLBM models is less than 3.6% up to this point. This concentration is not only far less than the maximum packing fraction for randomly dispersed spheres, it is below the percolation threshold of Gurland, but close to that of Bruggeman. Above this concentration the predictions of the Nielsen model begins to diverge from that of the SOLBM model. The Nielsen model predicts a higher relative thermal conductivity than the SOLBM model as the concentration of filler particles approaches the maximum packing fraction. Nielsen's curve assumed a maximum packing fraction of 0.637. Both theoretical curves used a value of 1000 for the ratio of the thermal conductivity of the filler to that of the matrix. As will be shown later, this ratio has little effect on the composite thermal conductivity when it is above 100:1, a ratio that includes all of the metals encompassed in the various studies.

Also contained in Fig. 4 is experimental data taken from a wide range of sources showing that all of these models are quite good up to a volume fraction

Table 2. Maximum packing fractions of common filler shapes

Filler	Shape	Packing	ϕ_{max}
generic	spheres	hexagonal	0.7405
generic	spheres	face centered cubic	0.7405
generic	spheres	body centered cubic	0.60
generic	spheres	simple cubic	0.524
generic	spheres	random loose	0.601
generic	spheres	random close	0.637
generic	irregular	random close	~ 0.637
talc	flakes	random 3D	0.40–0.56
mica	flakes	random 3D	0.38–0.45
Wollastonite	fibers	random 3D	0.62
asbestos	fibers	random 3D	0.60
glass/carbon	fibers	random 3D, $1/D = 10$	0.42
generic	spheres	bimodal[a], $d_2/d_1 = 7$	0.85
generic	spheres	broad	0.70

[a] Content ratio is five parts d_2 to one part d_1

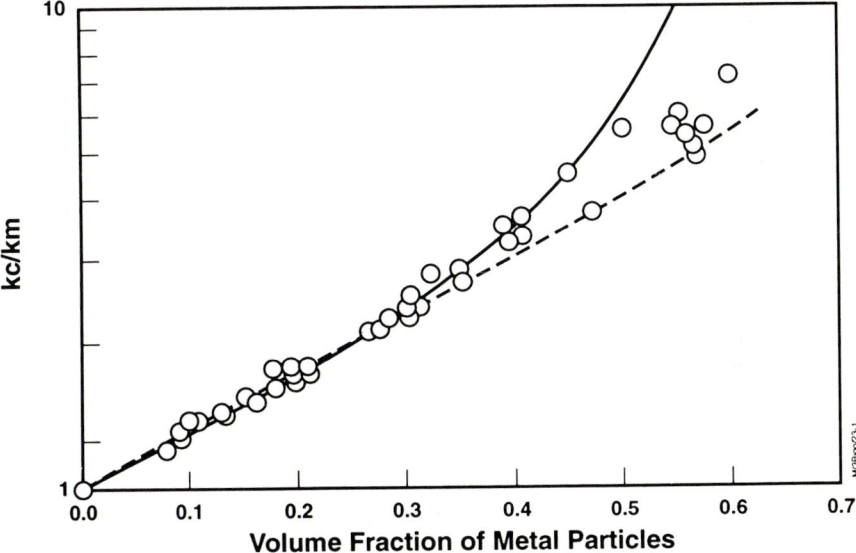

Fig. 4. Second order bound predictions of the relative thermal conductivity of a two phase system as a function of volume fraction of the dispersed phase in which the particles are spherical and $k_f/k_m \geq 1000$; Nielsen (—), and Hashin & Shtrikman, Hamilton & Crosser, Hatta & Taya (– –). Data taken from [17, 40–45]

of 0.30. This is the same data shown in Fig. 2. Note the significant improvement in the ability of the second order models to describe the data over the FOLBM. The data do not distinguish cleanly between Nielsen's model and the SOLBM over most of the range of data examined, as individual data points fall between the two curves. Nielsen's model does appear to overestimate the increase in thermal conductivity at concentrations very close to the maximum packing fraction. The spherical particles in the 12 composite systems plotted on this graph ranged from 2 µm to 12 mm, a very wide range of particles sizes. Five different kinds of metal spheres were included in the data. The data did not show any effects of either particle size or metal type on the relative increase in thermal conductivity. The only parameter affecting the ratio of k_c/k_m is the concentration of metal spheres in the composition.

The spread of the data in the high filler concentrations between the Nielsen model and the SOLBM results from the degree of formation, or lack of formation, of an interparticle network within the matrix. The data clearly show that there is no significant degree of particle-particle interaction effects up to a volume loading of 0.30 for spherical particles. Above this level of concentration the extent and degree of particle-particle interaction is not simply determined, and there is evidence that partial interaction also occurs. This is shown by the number of data points lying in between both models. There is simply no way of determining a priori the microscale structure in a composition at high filler concentrations. Hence scatter is inherent in any compilation of data at high filler loadings. Third and fourth order models have been developed to provide more precise predictions at high filler concentrations, but they require information on the microstructure of the composite.

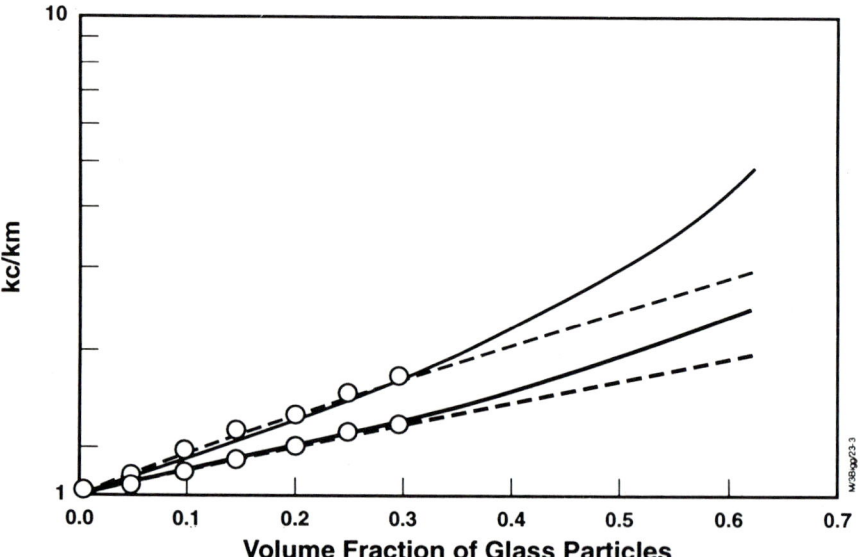

Fig. 5. Second order bound predictions of the relative thermal conductivity of a two phase system as a function of volume fraction of the dispersed phase in which the particles are spherical and k_f/k_m = 6.45; Nielsen (—), and Hashin & Shtrikman, Hamilton & Crosser, Hatta & Taya (– –), and k_f/k_m = 3.00; Nielsen (—), Hashin & Shtrikman, Hamilton & Crosser, and Hatta & Taya (– –). Data taken from [46]

Figure 5 contains data for spherical filler particles in polymer matrices in which the ratios of the thermal conductivity of the filler to that of the polymer is considerably less than 100:1. The filler material was the same in both instances (62 to 88 μm diameter glass spheres), but the matrix materials were different. Both the SOLBM and Nielsen's model accurately predict the behavior of these compositions, demonstrating their ability to account for the effect of the relative thermal conductivity of the filler to that of the matrix. In these cases, however, the maximum filler concentration was less than 30 vol%, so that network formation and packing considerations were not relevant.

Figure 6 shows the effect of filler thermal conductivity relative to matrix thermal conductivity over the range of ratios between 1 and 1000, as predicted by the SOLBM and the Nielsen model. As noted before, the predictions are essentially the same for filler volume fractions below 0.30. Above that filler concentration the predictions begin to diverge. This is shown in Fig. 6 where predictions for a volume fraction of 0.50 are also shown. Of interest is that both models show only a minimal effect of increasing filler thermal conductivity above a ratio of k_f/k_m greater than 100. This means that inorganic fillers such as CaO, MgO, and Al_2O_3 are as effective in increasing the thermal conductivity of polymers as metals. The practical consequence of this effect is that thermally conductive compositions can be produced without inducing electrical conductivity, a requirement in many appliance and circuit board applications.

Hashin and Shtrikman also provided an upper bound model for the thermal conductivity of spherical particles dispersed randomly in a continuous matrix

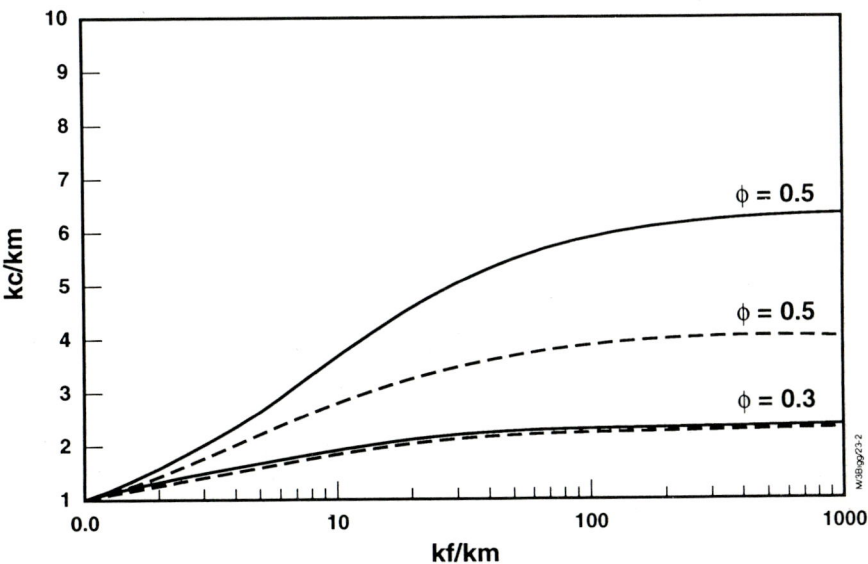

Fig. 6. The predicted effect of k_f/k_m for spherical filler particles on k_c/k_m according to Nielsen (—), and Hashin & Shtrikman, Hamilton & Crosser, and Hatta & Taya (– –) at filler volume fraction equal to 0.30 and 0.50.

[16]. That upper bound model is:

$$\frac{k_c}{k_m} = \frac{k_f[1 + (d-1)(1-\phi)b]}{k_m[1 - (1-\phi)b]} \tag{18}$$

where

$$b = \frac{k_m - k_f}{k_m + (d-1)k_f}. \tag{19}$$

This model is of use only when the relative thermal conductivities of the two phases are similar. In the common case, where the thermal conductivity of the filler is several orders of magnitude greater than that of the matrix, the predictions have little validity.

2.2 Irregularly Shaped Particles

Irregularly shaped particles are those which are non-spherical yet have no anisotropy. Many common fillers have this characteristic, including ground silica, clay, calcium carbonate, ground glass, wood flour, and alumina trihydrate. Modeling the effect of these fillers is often accomplished by altering a coefficient in a spherical model to account for the deviation from sphericity in the particles. In each of the SOLBM modes there is a coefficient that can be used to account for the effect of irregularly shaped particles. In Hashin and Shtrikman's model, Eq. (11), that coefficient is d. The adjustable coefficient in Hamilton and Crosser's model, Eq. (13), is N, while in Hatta and Taya's Eq. (14), the coefficient is S. In Nielsen's model it has already been pointed out that the coefficient A has different values for fillers with different shapes.

Figure 7 contains thermal conductivity data for composites produced from irregularly shaped particles having a ratio of k_f/k_m greater than 100. The particles, ranging from 10 μm to 0.60 mm, consisted of tin, aluminum, CaO, MgO, quartz, diamond, and Al_2O_3. As with the spheres, the dominant effect was the concentration of filler particles. Nielsen's model utilized a value of 3.5 for the coefficient A, while retaining a value of 0.637 for the maximum packing fraction. The coefficient S in Hatta and Taya's model was 0.25, and the parameter N in Hamilton and Crosser's model was 3, while Hashin and Shtrikman's model used a value of 4 for d. The SOLBM model appeared to provide a better fit to the data than Nielsen's throughout the entire range of filler loadings, although all three models provide satisfactory predictions up to filler loadings of 40 vol %. These results are quite interesting because, intuitively, data from composites filled with irregularly shaped filler particles should have provided more scatter than spherical fillers but the acquired data used in this plot do not have such a wide variation. This is considered serendipitous and not a reflection of being able to generate more precise data from irregular particles than spherical particles. Again, as with the composites filled with spherical particles, Nielsen's

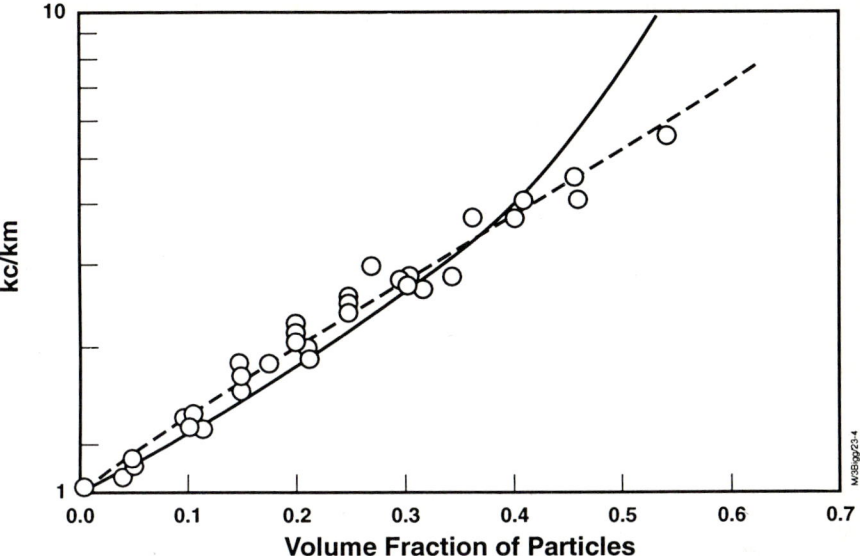

Fig. 7. Second order bound predictions of the relative thermal conductivity of a two phase system as a function of volume fraction of the dispersed phase in which the particles are irregular and $k_f/k_m \geq 100$; Nielsen (—), and Hashin & Shtrikman, Hamilton & Crosser, Hatta & Taya (- -). Data taken from [42, 46–48].

model appears to overestimate the thermal conductivity at the highest filler loadings.

Agari presented an approach to predicting the thermal conductivity of generalized dispersions (including spherical and irregular particles and fibers) based on logarithmic averaging of the thermal conductivity of the two components [21–25]. The logarithmic equation Agari and co-workers started with was:

$$k_c^n = \phi_m(C_1 k_m)^n + \phi_f k_f^{nC}. \tag{20}$$

In this expression, C_1 is a term attributed to the effect of the filler on the crystallinity and hence, thermal conductivity of the polymer matrix. Since k_m can be measured as a function of polymer crystallinity, and the effect of filler type and concentration on crystallinity can be measured independently of thermal conductivity measurements, the refinement of incorporating the term C_1 is unnecessary. Agari determined C_1 to be very close to 1 for most of the compositions studied by curve fitting Eq. (20) to data. Making $C_1 = 1$ simplifies Eq. (20), resulting in:

$$(k_c/k_m)^n = \phi_m + \phi_f(k_f/k_m)^{Cn}k_m^{n(C-1)}. \tag{21}$$

n is an index related to the degree of uniformity in the dispersion. In a perfect dispersion, n = zero, which reduces Eq. (21) to the trivial $\phi_m + \phi_f = 1$. At this point Agari assumed that n was not equal to zero, but close enough that the following approximations can be used:

$$(k_c/k_m)^n = 1 + n \log(k_c/k_m) \tag{22}$$

$$(k_f/k_m)^{Cn} = 1 + Cn \log(k_f/k_m). \tag{23}$$

Using this approximation, Eq. (21) becomes:

$$\log(k_c/k_m) = \phi_f C \log(k_f/k_m). \tag{24}$$

In order to make the coefficient C more than a curve fitting parameter having no predictive manner, it was related the network-forming ability of the particles. Whereas Nielsen used the maximum packing fraction as the key index of geometric packing, Agari used the volume concentration at which the particles formed a continuous network as the key particle structural parameter. Previous work showed that these two parameters are closely related, although an exact numerical relationship has never been determined [26]. The volume concentration at which a continuous particle network is formed is easy to determine for electrically conductive fillers, but not so easy for nonconducting fillers. Nielsen's choice of the maximum packing fraction would be a more appropriate parameter to relate to C. At the present time, however, C remains a curve fitting parameter for all except electrically conducting composites. The relationship between C and the critical volume fraction, ϕ_c, for network formation was:

$$C = \log(1/\phi_c). \tag{25}$$

ϕ_c is determined as the volume concentration at the inflection point in the resistivity vs filler concentration curve, a curve that undergoes a dramatic change of 10 to 15 orders of magnitude over a narrow range of filler concentration [26].

2.3 Flakes

Many filler particles are in the shape of flakes, particles whose basic geometry is that of a disc or plate. These particles have a characteristic aspect ratio of some length parameter divided by thickness. Common flake filler particles are mica and talc. Metal flakes have also been investigated as a means of providing electrical conductivity to polymers. Hatta and Taya developed a model to describe the thermal conductivity of a composite filled with flakes [27]. That model uses the same basic equation they developed for spherical and irregular filler particles:

$$\frac{k_c}{k_m} = 1 + \frac{\phi}{S(1 - \phi) + k_m/(k_f - k_m)}. \tag{26}$$

There is one significant difference, however. Since flakes have a characteristic geometry, the effect of particle orientation must be taken into account. If the flakes are aligned along the plane of a sheet of the composite, S depends on whether the thermal conductivity was measured along the plane or through the thickness of the sheet, perpendicular to the orientation of the flakes. When measured along the plane of the flakes:

$$S = \pi L/4X \tag{27}$$

where L is the effective diameter of the flake and X is the thickness of the flake. The value of S when the measurement is made perpendicular to the plane of the flakes is:

$$S = 1 - \pi L/2X. \tag{28}$$

Hatta and Taya mention that their theoretical predictions were below experimental values for at least one system of flake filled compositions. They mentioned the possibility of a change in the material as the cause for the deviation. More likely, it is the failure to account for the fact that flakes will rarely be perfectly aligned within a plane. While there is usually some preferential alignment, there is also a significant quantity of flakes that are out-of-plane. If the flakes are very small in diameter, as talc particles often are, the flakes may even be distributed in a random manner. In that case the composite should be treated as a randomly dispersed isotropic composition, and Eqs. (11), (13) and (14) used with appropriately modified coefficients. Overall there is little data available on composites filled with aligned flakes. In practice it is difficult to produce flake filled composites with an aligned but randomly dispersed in-plane filler arrangement.

2.4 Short Fibers

The development for fibers is even more complex. Hatta and Taya presented three similar equations to describe randomly oriented: isotropic short-fiber reinforced composites, through the plane aligned but randomly oriented long-fiber reinforced composites, and in-plane aligned long-fiber reinforced composites [28]. The equations used to predict the thermal conductivity of an isotropic short-fiber reinforced composite are:

$$k_c/k_m = 1 + \phi[(k_f - k_m)(2S_{33} + S_{11}) + 3k_m]/J \tag{29}$$

$$J = 3(1 - \phi)(k_f - k_m)S_{11}S_{33} + k_m[3(S_{11} + S_{33}) - \phi(2S_{11} + S_{33})] + 3k_m^2/(k_f - k_m) \tag{30}$$

$$S_{11} = \frac{1/D}{2[(1/D)^2 - 1]^{3/2}}\{(1/D)[(1/D)^2 - 1]^{1/2} - \cosh^{-1}(1/D)\} \tag{31}$$

$$S_{33} = 1 - 2S_{11}. \tag{32}$$

In these equations l is the average fiber length, and D is the fiber diameter. The thermal conductivity of planar composites in which the fibers are randomly arranged within the plane and aligned along the plane the thermal conductivity in the planar direction is:

$$k_c/k_m = 1 + \phi(k_f - k_m)[(k_f - k_m)(S_{33} + S_{11}) + 2k_m]/H \quad (33)$$

$$H = 2(1 - \phi)(k_f - k_m)^2 S_{11} S_{33} + k_m[(k_f - k_m) \\ \times (2 - \phi)(S_{11} + S_{33}) + 2k_m^2. \quad (34)$$

The thermal conductivity of such a composite measured through the thickness of the plane is given by Eq. (26) in which the appropriate value of S is $1 - 2S_{11}$.

Nielsen's model was not developed to predict the behavior of fiber reinforced composites, except for the case of isotropic, very short fiber reinforced composites [3]. It cannot adequately predict the thermal conductivity of composites measured through the plane when the fibers are randomly dispersed within the plane.

Agari used his logarithmic model to accurately predict the thermal conductivity of a number of carbon fiber filled polyethylene composites [24]. It was demonstrated by multidimensional electrical measurements that the composites prepared were isotropic. With the composites prepared in an isotropic manner, the coefficient C was determined from electrical resistivity measurements (by using Eq. (25)). The maximum degree of thermal conductivity increase for the composites investigated was by a factor of 3.25. This composite was filled with 30 vol% carbon fibers that had an aspect ratio of 13.4. Using this data, Hatta and Taya predict a k_c/k_m of 3.15, while Nielsen's model predicts a value of 6.6. Clearly the Hatta and Taya model does an excellent job of predicting the thermal conductivity of the isotropic, short fiber, reinforced composite. Nielsen's model does not accurately predict the measured value. It predicts a higher value than measured due to the fact that this model gives greater weight to the dispersed phase than can be experimentally justified at filler concentrations near the maximum packing fraction. This overestimate in composite thermal conductivity at very high filler loadings was previously pointed out for spherical and irregularly shaped particles. Fibers with an aspect ratio of 13.4 have a maximum packing fraction of 0.31. This was arrived at from Fig. 8, which shows the relationship between fiber aspect ratio and the maximum packing fraction for isotropically filled, randomly dispersed composites [27].

There is little available data to analyze the predictive ability of these models to evaluate the through-the-plane conductivity of composites in which the fibers are randomly oriented within a plane. In Delmonte's text there is limited data on the thermal conductivity of carbon fiber reinforced polyamide 66 composites, measured through the thickness of the plane of orientation of the fibers [26]. This data is summarized in Table 3. The fiber aspect ratios were taken from Fig. 8, based on the assumption that sufficient 3-D mixing occurred in the compounding and injection molding machines to reduce the fibers to an aspect ratio compatible with the maximum fraction associated with a random isotropic condition. The Hatta and Taya model does quite well in predicting the thermal conductivity across the plane of fiber orientation for these composites.

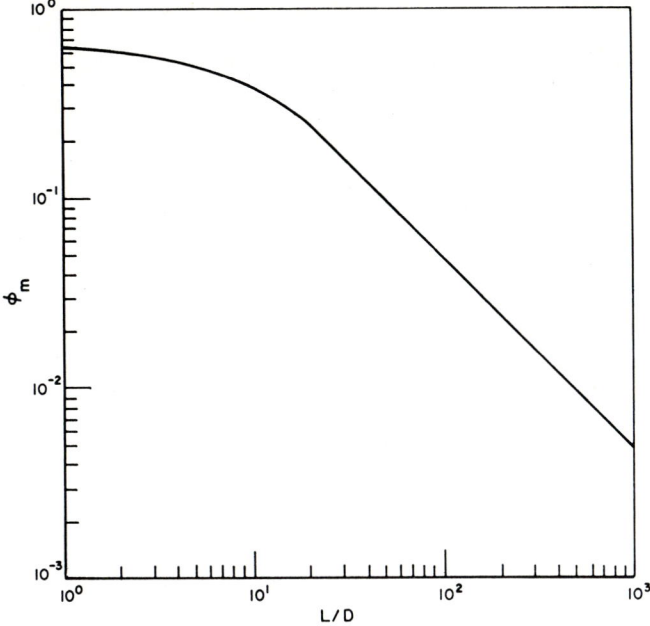

Fig. 8. Maximum packing fraction vs fiber aspect ratio for random isotropic dispersions

Table 3. Through-the plane thermal conductivity of carbon fiber reinforced polyamide 66 composites

L/D	ϕ	k_m cal/cm-s-C	k_f cal/cm-s-C	k_c/k_m Hatta & Taya Eq. 26	k_c/k_m Experimental
35	0.13	0.00058	0.00995	3.02	3.2
25	0.20	0.00058	0.00995	4.05	4.1
17	0.28	0.00058	0.00995	5.10	5.1

It would be appropriate to present data verifying the theoretical predictions for a wide variety of conductive fiber compositions. Unfortunately, few data are available on such systems. It has been shown that Neilsen's theoretical development cannot be used to predict the thermal conductivity of polymers containing long fibers because it cannot account for the variety of possible fiber distributions in an injection molded or extruded fiber reinforced thermoplastic composite [3].

2.5 Long Fibers

The thermal conductivity of long fiber reinforced composites should be somewhat easier to model than short fiber reinforced composites because the fibers are more likely to be aligned along the plane of a sheet than short fibers. Short fibers have a tendency to random alignment or localized preferential orientation. Therefore the equations developed by Hatta and Taya for in-the-plane and through-the-plane fiber reinforced composites should be useful. Unfortunately, there is very little data on the thermal conductivity of long fiber reinforced composites. The limited data shown in Table 3 for short fiber reinforced composites does provide a degree of confidence in using this model.

2.6 Continuous fibers

Continuous fiber reinforced composites are modeled by the first order upper and lower bound models. The upper bound model accurately describes the thermal conductivity along the fiber direction, while the lower bound model describes the thermal conductivity in the direction perpendicular to the fibers. In

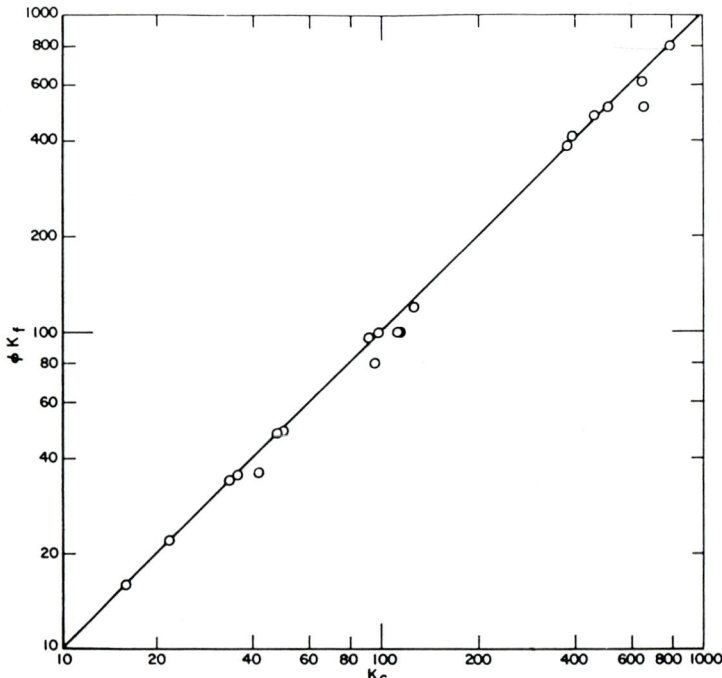

Fig. 9. Relationship between k_c calculated from the rule-of-mixtures (Eq. 1) and ϕk_f. Data taken from [49–51].

these cases the model actually describes the physical system, and should be expected to provide an accurate description of the behavior of the material. This is shown clearly in Fig. 9 for a series of unidirectionally oriented carbon fiber reinforced composites measured in the fiber direction [3]. Since the product of k_m and ϕ_m is several orders of magnitude less than $k_f \phi_f$, only the dominant term is plotted in this figure. Carbon fibers are interesting from a thermal transport point-of-view, since their thermal conductivity along the fiber axis is several orders of magnitude higher than in the transverse direction.

3 Measurement Techniques

There are two basic approaches to measuring the thermal conductivity of materials. The most basic approach is a steady state approach based on Fourier's definition of thermal conductivity. Steady state measurements often require a long time to reach equilibrium and, as a result, several unsteady state approaches have been developed that are based on the rate of change of temperature of a material suddenly subjected to a new thermal environment.

3.1 Steady State Techniques

The most basic techniques for measuring thermal conductivity is based on Fourier's definition of thermal conductivity, as presented in Eq. (1). A simple hot plate device, as shown in Fig. 10, is often used. In such a device heat is generated steadily by a source; an electrical source is shown in Fig. 10. The heat flow is

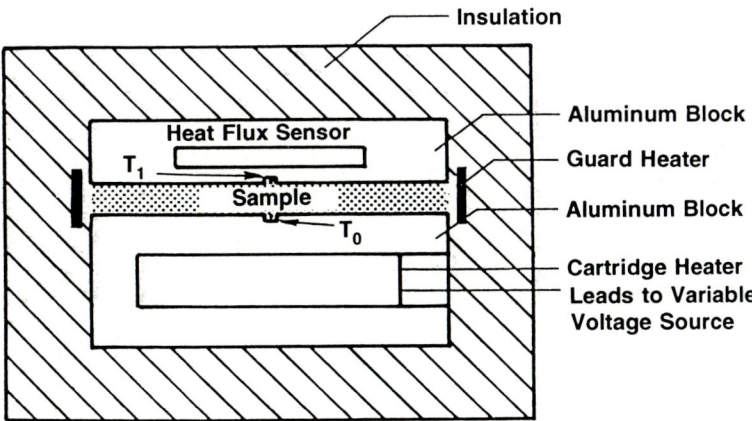

Fig. 10. Schematic of Guarded Hot Plate device used for steady state measurement of thermal conductivity

directed through a flat sample that is confined between two parallel platens. The rate of heat flow through the sample is measured by a flux sensor. Thermocouples located at the sample-fixture interfaces record the temperature drop across the sample. The entire unit is insulated to minimize thermal losses and to reduce the time required for the system to reach equilibrium, a time which is often several hours. A heated guard is often placed around the edges of the sample to eliminate heat loss through the edges of the sample. This is required if the sample is thick. Thick samples are often required to produce a reliable temperature drop across the sample. The guard is maintained at the average temperature of the sample. Sources of error are primarily related to imperfect sample-fixture contact at the interface. Interfacial contact is often improved by adding a thin thermal grease coating to the surfaces of the sample to eliminate areas of noncontact. An alternative method is to evaluate samples of different thickness, and then subtract the interfacial resistance from the overall thermal resistance to determine only the resistance due to the thermal conductivity of the material [29].

More sophisticated flat plate devices utilize a boiling liquid as the heat source directly against the lower face of the sample, and a condensing liquid at the opposite face. Since the heats of vaporization and condensation are known, as are the temperatures of vaporization and condensation, all the requirements of Eq. (1) are met directly without interfacial problems. The rate of vaporization and condensation must be measured in these units to provide the rate of heat transfer.

Techniques based on Fourier's law of heat transfer, Eq. (1), are not adept at handling molten polymers nor the effect of pressure. Molten polymers often degrade during the time required to reach equilibrium. Molten polymers cannot be used if the heat source is the gas from a boiling liquid. While pressure can be applied when the sample is contained between two flat plates, most commercial units are not equipped to handle significant levels of pressure.

3.2 Unsteady State Approaches

Unsteady state techniques have been developed to reduce the time required to generate data. The basic equation for which most unsteady state measurement techniques are based is [30]:

$$\frac{\delta^2 T}{\delta x^2} = \frac{\rho C_p}{k} \frac{\delta T}{\delta t} \tag{35}$$

where ρ is the density of the material and C_p is the specific heat; t is time. The solution to this equation depends on the geometry of the system. The most common system is the line source technique (31). This approach consists of using a resistive wire embedded through the center of a sample that is commonly of cylindrical shape. A step function of electrical power, constant current and voltage, is passed through the wire to provide a constant level of energy dissipation. The temperature is recorded as a function of time in the sample at some distance, r, away from the heated wire. Neglecting end effects and assuming that the term $r^2 C_p/4kT$ is small, Eq. (35) can be solved and expressed as

$$k = \frac{q'}{4\pi(T_2 - T_1)} \ln(t_2/t_1) \tag{36}$$

where q' is the rate of heat generation per unit length along the wire. T_1 is the temperature at time t_1, and T_2 is the temperature at time t_2. Such a system can be pressurized and can handle polymers that are molten from the wire to the thermocouple. The time required to generate a reliable measurement is of the order of 15–30 s with this technique [32].

The principal difficulty associated with this technique is sample preparation. Existing molded samples often cannot be used. Placement of the thermocouple can also be a problem in molded samples. Cast samples are usually easier to prepare than molded samples. With cast samples, however, the issue of composite homogeneity must be considered.

Since cylindrical samples must be specially prepared, and are often different from samples prepared by production techniques, a modification of the line source technique has been developed that allows for the use of a rectangular sample [31]. In this method the wire source is replaced by a thin metal foil. If the sample is sufficiently thick that it can be considered infinitely thick, Eq. (35) can be solved for unsteady state heat flow into an infinitely thick heat sink as follows [30]:

$$T = T_o + \frac{2Q}{Ak}\sqrt{kT/\rho C_p}\,\mathrm{ierfc}(x/(\sqrt{4kT/\rho C_p}). \tag{37}$$

Ierfc(z) is the integral complementary error function which is tabulated in standard mathematical references. T_o is the initial temperature of the sample prior to being brought into contact with a heat source that provides a constant rate of heat, Q. k can be determined by measuring the sample temperature at a given distance from the heat source, x, at a given time, t. This procedure requires intimate contact between the heat source and sample, as well as the placement of the thermocouple at a location which meets the assumption of a plate that is infinitely deep. In practice two samples are used – one sample on either side of the foil. Half of the heat generated goes into each sample.

Analogous to this approach is a technique in which the constant energy source is replaced by a constant temperature surface, T_i. Boiling liquids are often used to provide a constant surface temperature in addition to minimizing interfacial thermal resistance. In utilizing this technique the appropriate solution to Eq. (35) is [30]:

$$T = T_o + (T_o - T_i)\mathrm{erf}(x/\sqrt{4kt/\rho C_p}). \tag{38}$$

Erf(z) is the error function, which is also tabulated in most standard mathematical texts.

Unsteady state thermal conductivity measurements can also be made with an oscillating energy source. Dynamic thermal measurements have been developed by Cahill and Pohl, Birge and Nagel, and Frank, Drach, and Fricke [33–35]. The techniques developed by these groups utilize the basic line (or thin film) source geometry, but instead of passing constant electrical energy through the heating resistor, the current is passed through the heating element in an oscillatory manner at a frequency ω. Measurements of the oscillating thermal response are then used to determine the thermal conductivity of the material. The current carrier is a vapor deposited metal strip of film condensed directly onto the sample to be measured. This insures intimate contact between the energy source and sample, thus minimizing contact resistance. The strip has a width that is small compared to the wavelength of the diffusive thermal wave.

Electrical current is passed through the metal strip in a sinusoidal wave at an angular frequency, ω, with a constant line voltage, V. This generates a thermal wave at the frequency 2ω. The metal strip also acts as a thermal sensor, which produces a voltage oscillation at the frequency 3ω. This technique is referred to in the literature as the 3ω technique for that reason. A spectrum analyzer is used to detect the harmonic voltage signal at 3ω. By making measurements at two frequencies the thermal conductivity of a material can be determined according to the following equation:

$$k = \frac{V^3 \ln(\omega_2/\omega_1)}{4\pi/R^2(V_{3,1} - V_{3,2})} \frac{dR}{dT} \qquad (39)$$

where V is the average rms line voltage at frequency ω, R is the average resistance at the voltage and frequency ω. l is the length of the metal strip. ω_1 is the first frequency, ω_2 is the second frequency, and $V_{3,1}$ and $V_{3,2}$ are the harmonic rms voltages at $3\omega_1$ and $3\omega_2$ respectively. dR/dT is the slope of wire resistance as a function of temperature. This slope is generated separately in a calibration run.

This technique has the advantage of being fast and measurements are made in seconds. Measurements can also be made on small samples. The technique is difficult to use on composites if the thickness of the metal strip is of the order of the size of the filler particles. Since the metal strip is vacuum deposited onto the surface of the sample, it is possible to generate data on non-flat samples.

4 Applications

As mentioned in the Introduction there are many reasons for generating thermal conductivity on polymer composites. These include increased accuracy in modeling molding and extrusion processes, the design of circuit boards, and the design and selection of heat exchange devices. One of the most important application areas where reliable thermal conductivity data are needed is the area of polymeric heat exchangers, where the thermal conductivity directly effects the performance of the device [36, 37]. Low cost metals corrode rapidly in heat exchangers in which corrosive chemicals are processed. Certain alloys have been identified that are corrosion resistant, but at a considerable cost penalty. Consequently, corrosion resistant polymeric materials have been considered possible alternatives to metals in heat exchangers. As discussed previously, increases in the thermal conductivity of an isotropically molded plastic composite were found to be practically limited to approximately ten times that of the unfilled polymer. For spherical and dimensionally isotropic irregularly shaped filler particles, the influence of increasing filler conductivity was negligible when the ratio of filler conductivity to matrix exceeded 100:1. For fibrous fillers, the upper limit of composite thermal conductivity was limited by the packing geometry possible in the molded part. This effect became more pronounced as the aspect ratio of the fiber and concentration continued to increase. There are limits to the concentration of fibers that can be added to a polymer in a random manner that depend on the aspect ratio of the fibers. This was shown in Fig. 8 for isotropically filled compositions. This graph is a plot of maximum packing fraction vs fiber aspect ratio. It shows quite clearly that attempts to increase the fiber concentration result in reductions in fiber aspect ratio whenever there is a random three dimensional processing step undertaken. All instances in which higher fiber aspect ratios are obtained at volume concentrations greater than indicated in Fig. 8 involve some degree of fiber orientation, and loss of isotropy. The primary effect in plastics molding is to align the fibers in the planar direction of the molded product. This gives rise to the earlier consideration of random in-plane composites. While this helps to maintain fiber aspect ratio and composite strength, it does reduce the thermal conductivity in the perpendicular direction for some composites. There is relatively little effect when the reinforcement is a less conductive fiber such as glass fibers. But it is significant when the fibers are highly thermally conductive in the fiber direction, but considerably less so in the transverse direction, such as carbon fibers.

Figure 11 shows the theoretical relative heat transfer rate of a staggered finned tube heat exchanger as a function of the thermal conductivity of the tube material [36]. A heat exchanger with elements whose thermal conductivity is ten times that of a pure polymer performs at 95% of that expected from stainless steel tubes. Since this level of thermal conductivity increase is easily obtained with polymer composites, they represent viable candidates with regard to heat transfer. Mesloh has shown experimentally, in a forced convection heat exchanger, that an aluminum flake-filled plastic having a thermal conductivity 20 times that of the base polymer has an overall heat transfer rate of 75% of that of an aluminum tube [38].

In combined heat transfer situations, it was shown that a modest increase in conductivity of the plastic is sufficient to shift the controlling heat transfer factor from element thermal conductivity to the condensing film heat transfer coefficient [38]. In addition to having potentially adequate thermal conductance, certain plastics promote dropwise condensation in a condensing heat exchanger,

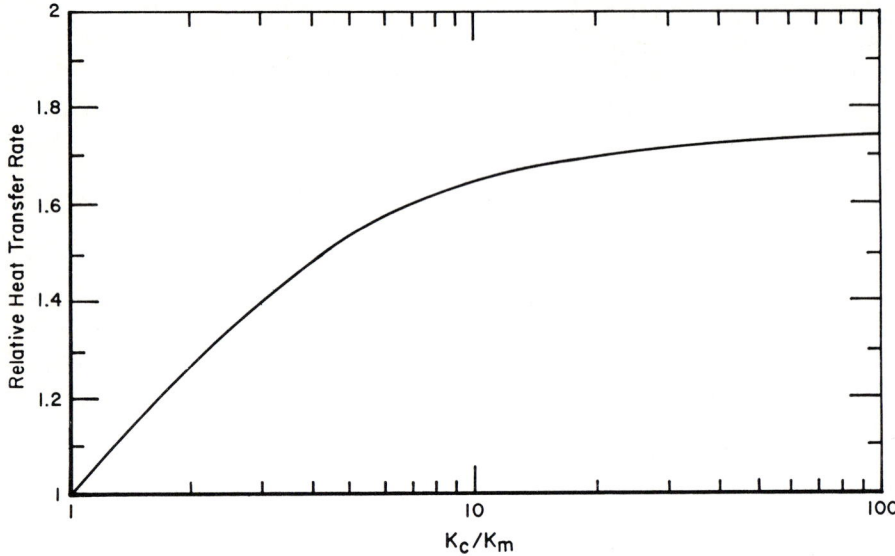

Fig. 11. Theoretical relative heat transfer rate of a staggered finned tube heat exchanger as a function of the relative thermal conductivity of the tube material

which is an even more efficient means of heat transfer than filmwise condensation [39]. This would serve to compensate for the reduction in heat transfer through the plastic wall.

5 Discussion

Several second order models are available for predicting the thermal conductivity of two phase, heterogeneous dispersions. These include the models of Hatta and Taya, Hashin and Shtrikman, Nielsen, and Agari. Agari's model is difficult to use for fillers which are not electrically conductive since a key parameter in the model is the volume concentration at which a continuous network of filler particles is formed. At filler loadings above this concentration the effect of filler packing has a strong influence on the thermal behavior of the composite. This accounts for the scatter in the data at high filler concentrations. It also accounts for the deviation among the various models above this filler concentration. This is shown in this text for both spherical fillers and irregularly shaped fillers. Although Nielsen's model accounts for packing considerations in a reasonable way, it appears to overestimate the thermal conductivity at concentrations near the maximum packing fraction. All of the models break down when the actual packing structure of the particles deviates from that assumed by the models.

Filler concentration and the thermal conductivity of the filler are the primary factors determining the thermal conductivity of a heterogeneous, randomly dispersed, two phase composite. When the ratio of the filler thermal conductivity to matrix conductivity exceeds 100:1 there is little further improvement in the thermal conductivity of the composite when the filler thermal conductivity is increased beyond this ratio. This means that highly conductive composites can be produced without inducing electrical conductivity by the use of such high conductivity insulating fillers as Al_2O_3, MgO, and quartz. The highest thermal conductivities can be achieved by selecting the filler particles such that they have either a broad or a bimodal particle size distribution such that very high loadings can be achieved. Loadings as high as 70 vol % have been achieved by such adjustments to the particle size distribution. Because spheres and non-directional irregularly shaped particles are capable of higher filler loadings than fibers, such composites are capable of producing composites with higher thermal conductivities than fiber filled composites. Under ideal conditions the thermal conductivity of a composite can be as high as 20 times that of the base polymer.

6 References

1. Mottram JT (1992) Material Design 13:221
2. Progelhof RC, Throne JL, Reutsch RR (1976) Polym Engr Sci 16:615
3. Bigg DM (1986) Polym Comp 7:125
4. Torquado S (1987) Reviews in Chem Engr 4 (3&4):151
5. Mottram JT, Taylor R (1991) In: Lee M (ed) International encyclopedia of composites, vol 5, VCH, New York, p 476
6. Segre G, Silberberg A (1962) J Fluid Mech 14:115
7. Karnis A, Goldsmith HL, Mason SG (1961) J Colloid Interfacial Sci 22:531
8. Sahimi M, Scriven LE, Davis HT (1984) J Phys C: Solid State Phys 17:1941
9. Kutcherov V, Chernoutsan A (1993) J Appl Phys 73(5):2259
10. Shah N, Ottino JM (1986) Chem Engr Sci, 41:283
11. Bruggeman DAG (1935) Ann Phys 24:636
12. Gurland J (1966) Trans Met Soc AIME, 236:6423
13. Bigg DM (1984) Adv Polym Tech 4:255
14. Torquado S (1985) J Appl Phys 58:3790
15. Torquado S (1984) J Chem Phys 81:5079
16. Hashin Z, Shtrikman S (1962) J Appl Phys 33:1514
17. Hamilton RL, Crosser OK (1962) Ind Engr Chem Fund 1:187
18: Hatta H, Taya M (1985) J Appl Phys 58(7):2478
19. Eshelby JD (1957) Proc R Soc London A241:376
20. Nielsen LE (1974) Ind Engr Chem Fund 13:17
21. Agari Y, Uno T (1986) Jour Appl Polym Sci 32:5705
22. Agari Y, Tanaka M, Nagai S (1987) Jour Appl Polym Sci 34:1429
23. Agari Y, Ueda A, Tanaka M, Nagai S (1990) Jour Appl Polym Sci 40:929
24. Agari Y, Ueda A, Nagai S (1991) Jour Appl Polym Sci 43:1117
25. Agari Y, Ueda A, Nagai S (1993) Jour Appl Polym Sci 49:1625
26. Delmonte J (1981) "Technology of Carbon and Graphite Fiber Composites", Van Nostrand Reinhold, New York
27. Hatta H, Taya M, Kulacki FA, Harder JF (1992) J Compos Mater 26:612

28. Bigg DM (1986) "Metal-Filled Polymers", ed. Bhattacharya SK, Marcel Dekker, New York, 615
29. Fletcher LS, Cerza MR, Boysen RL (1975) Progree Astronaut Aeronaut 49:371
30. Bird RB, Stewart WE, Lightfoot EN (1960) "Transport Phenomena", J Wiley & Sons New York
31. Hands D (1977) Rubber Chem Technol 50:480
32. Lobo H, Newman R (1990) "SPE Technical Papers", XXXVI 862
33. Cahill DG, Pohl RO (1987) Phys Rev B 35(8):4067
34. Birge NO, Nagel SR (1985) Phys Rev Lett 54:2674
35. Frank R, Drach V, Fricke J (1993) Rev Sci Instrum 64:760
36. Bigg DM, Stickford GH, Talbert SG (1989) Polym Engr Sci 29:1111
37. Sullivan HF, Wright JL (1982) "Symposium on Condensing Heat Exchangers", Atlanta GA (Mar 3–4)
38. Mesloh RE (1986) "Industrial Heat Exchanger", (ed. Hayes AJ, Liang WW, Richlen SL, Tabb ES), ASM Publications New York
39. Paschke LF (1984) Chem Engr Prog 80:70
40. Nieberlein VA (1978) IEEE Trans Comp Hybrid Manuf Tech CHMT-1 172
41. Nieberlein VA, Steverding B (1977) J Mater Sci 12:1685
42. de Araujo FFT, Rosenberg HM (1976) J Phys D: Appl Phys 9:665
43. de Araujo FFT, Garrett KW, Rosenberg HM (1976) ICCM Proc Intl Conf Compos Mater 2:568
44. de Araujo FFT, Rosenberg HM (1976) NTIS AD-A024963
45. de Vera AL, Streider W (1977) J Phys Chem 81:1783
46. Sundstrom DW, Lee Y (1972) J Appl Polym Sci 16:3159
47. Garrett KW, Rosenberg HM (1974) J Phys D: Appl Phys 7:1247
48. Katz HS, Milewski JV (1978) "Handbook of Fillers and Reinforcements", Van Nostrand Reinhold, New York
49. Kalnin IL (1974) ASTM Spec Publ Symp Compos Reliab 580:560
50. Rosenberg HM (1977) NTIS AD-A040950
51. Knibbs RH, Baker DJ, Rhodes G (1971) SPI Ann Conf Reinf Plast 26:8-F

Editor: Prof. Godovsky
Received May 1994

Model Treatments of the Heat Conductivity of Heterogeneous Polymers

V. P. Privalko and V. V. Novikov
Institute of Macromolecular Chemistry, Academy of Sciences of Ukraine, 253160, Kiev, Ukraine

Basic physical concepts and limitations of current approaches to the theoretical description of the composition dependence of heat conductivity of microheterogeneous polymer materials (MHM) are reviewed. All "pragmatic" approaches (i.e., those assuming the existence of a infinitely thin, "mathematical" interface between the components) fail to account explicitly for salient structural features of MHM such as the onset of an "infinite" cluster of a disperse component at the percolation threshold, and the transition of a portion of a continuous component into a structurally different "boundary interphase" (BI). Among the "physical" approaches, it is apparently the Step-by-Step-Averaging (SSA) model which accounts simultaneously for both cited structural features of MHM. The SSA model was shown to provide a quantitative description of the experimental data available by an appropriate choice of relevant BI parameters (i.e., thickness and "partial" heat conductivity): at the present stage, however, the numerical values of the latter should be considered as fitting variables, rather than true material properties of BI.

1	Introduction	33
2	Theoretical Aspects: Pragmatic Approaches	34
	2.1 Closure of Equations for Effective Conductivity	34
	2.2 Method of Direct Calculations	37
	2.3 Self-Consistent Field Approach	38
	2.4 Method of Integration	39
	2.5 Method of Averaging over the Ensemble	40
	2.6 Lichtenecker's Approach	41
	2.7 Herring's Approach	41
	2.8 Correlation Approximation. Method of Conditional Moments .	42
	2.9 Variational Approach	42
	2.10 Method of Functional Inequalities	45
	2.11 Method of Integral Sectioning	47
3	Theoretical Aspects: Physical Approaches	52
	3.1 Step-by-Step Averaging Approach	52
	3.1.1 The Percolation Model	52
	3.1.2 Voronoi Polyhedra. The Model of Equivalent Element . .	58

4 Tests of Model Predictions. 63
   4.1 Analytical Test. 63
   4.2 Experimental Test. 66
       4.2.1 Filled Polymers 66
            4.2.1.1 Filled Cross-Linked Epoxies 66
            4.2.1.2 Polyethylene (PE)/NaCl 67
            4.2.1.3 Polyethylene (PE)/AgCl 70
            4.2.1.4 Polyethylene (PE)/Quartz; PE/Al_2O_3;
                       PE/Graphite. 70
       4.2.2 Polymer Blends 72

5 Conclusions. 73

6 References 74

1 Introduction

Basically, the whole business of binary and/or multicomponent polymer materials started from the intuitively obvious idea that literally any property P of such a material may be varied within the limits set up by the corresponding properties of individual components 1 and 2 (i.e., P_1 and P_2). A proper choice of a composite material with desired properties requires, however, knowledge of the pattern of a property-composition dependence which can be derived from consideration of the relevant theoretical model. Perhaps the simplest model of a two-component polymer MHM would be a homogeneous dispersion of isolated particles of component 1 (e.g., filler) in a structureless, continuous medium 2 (polymer) [1]. The early theories based on these concepts (hereafter referred to as "pragmatic" approaches) were moderately successful only at a fairly low volume content of the disperse phase, φ. As an immediate guess, failure of such theories at higher φs could be tentatively attributed to the neglect of a possible change of a mother phase 2 as a result of its strong interfacial interactions with the disperse phase 1 ("interfacial effect"). Moreover, it was also recognized that, keeping φ constant, the effective property P may vary from the upper bound, as in the case of a "soft inclusions in a continuous rigid matrix"-type morphology, to the lower bound, as in the inverse case of a "rigid inclusions in a continuous soft matrix"-type one ("morphological effect"). These ideas, essentially, were implicit in the more up-to-date theories based on the "self-consistent field" (SCF) or "effective medium" (EM) approaches, according to which any effective property of a continuous matrix 2 in presence of a disperse phase 1 should be the same as that sought for a binary composite.

These latter approaches, although having somewhat improved the situation, still proved inadequate as far as a quantitative description of the property evolution over the whole range of φ is concerned. In the course of a continuous accumulation of a larger body of experimental data, it was becoming more and more evident that any structural model of polymer MHM claiming validity must account explicitly for the following physical phenomena.

i) All the former "pragmatic" approaches assumed the existence of a sharp, "mathematical" (i.e., infinitely thin) interface between the components. The invalidity of this assumption was proved by experiments (e.g. [2]) according to which the interactions between the component of whatever polymer MHM (filled polymers, phase-separated polymer blends, block copolymers, and the like) invariably resulted in a smearing-out of a "mathematical" interface into a "physical" interfacial region. However, it had to be recognized that the influence (if any) of a (presumably, high surface energy) disperse phase 1 on a continuous matrix 2 may extend not to any geometrically possible distance between the particles as implicit in SCF approach, but is limited to a smaller scale characteristic for interfacial interactions which are known to decay quickly with distance from the phase boundary [3]. In other words, one has to introduce the concept of a special "boundary interphase" (BI) [2–4], the thickness of which

Δr may be defined as a characteristic distance at which the studied property of BI becomes indistinguishable from that of a pure matrix phase 2 [5].

ii) In the course of a continuous increase of φ, the particles of a disperse phase 1 will not remain isolated and distributed randomly, but would tend to aggregate into clusters up to the critical volume content φ_c ("percolation threshold" [6, 7]) at which all the disperse particles become involved in an 'infinite cluster' (InC) spanning the whole volume of MHM. This means, in effect, that the composition interval around φ_c corresponds to the coexistence of two 'interpenetrating', continuous phases 1 and 2. Thus those theoretical approaches which explicitly take into account both items i) and ii) will be referred to as "physical" approaches.

The present review is accordingly divided into two parts, the first outlining the basic features of "pragmatic" approaches, and the second dealing with their "physical" counterparts. It is to be understood that, in view of the abundance of scientific literature with papers backing either of the two cited approaches, reference will be made only to those papers which are presumably representative as concerns the evolution of the main physical ideas behind the mathematical formalism.

2 Theoretical Aspects: Pragmatic Approaches

2.1 Closure of Equations for Effective Conductivity

The understanding of the effective properties of MHM implies the knowledge of the pattern of physical fields distribution in all components of MHM from which the properties of interest may be derived using the approximation of a quasihomogeneous medium [8–11]. The first step involves specification of the 'bulk representative elements' (BRE) which completely fill the total volume V and which have the same properties as MHM. It is assumed that the volume occupied by BRE is statistically homogeneous (i.e., the volume content of components, two-point correlations and other properties of BRE are invariant with respect to their location in the given volume V) as soon as this would conform to the ergodicity criterion (i.e., equality of the averages over the statistical ensemble and over the volume).

The statistically homogeneous field in MHM may be generated by a proper choice of boundary conditions; in the case of heat conductivity problems, the latter may be defined as

$$t(s) = \langle \nabla t \rangle \mathbf{r}, \quad q_n(s) = \langle \mathbf{q} \rangle \mathbf{n}, \tag{1}$$

where t(s) is the temperature on the surfaces surrounding the volume V, $q_n(s)$ is the normal heat flux to the surface, **r** is the vector-radius, **n** is the vector-normal to the surface, and $\langle \nabla t \rangle$ and $\langle q \rangle$ are the temperature gradient and the heat flux averaged over volume V, respectively.

The effective heat conductivity λ and resistivity ρ are defined by Eqs. (2):

$$\langle q \rangle = -\lambda \langle \nabla t \rangle; \tag{2.a}$$

$$\langle \nabla t \rangle = -\rho \langle q \rangle, \tag{2.b}$$

where $\lambda \rho = 1$. In a similar fashion, one may write for the local regions of MHM (e.g., components)

$$q(r) = -\lambda(r)\nabla t(r); \quad \nabla t(r) = -\rho(r)q(r), \tag{3}$$

where $q(r)$, $\nabla t(r)$, $\lambda(r)$ and $\rho(r)$ are the random functions of coordinates.

Equations (2) and (3) also apply to several other properties (e.g., electrical conductivity, dielectric permittivity, magnetic susceptibility, diffusivity, etc.) which may all be united under a common generic name, generalized conductivity.

Since the properties of a homogeneous material are usually assumed to be invariant with respect to its dimensions down to the infinitesimal differential volumes, the field equations may be written in derivatives. However, as far as any real material has its intrinsic microstructure, the appropriate quantity in this case would be the 'bulk differential element' (BDE) which is assumed to consist of a sufficiently large number of small (compared to both the total volume V and BREs) microelements ("crystals"). In the case of MHM, the usual differential field equations of generalized conductivity are assumed to apply; i.e., the mean value of the function $f(r)$ in the point r is defined as

$$\langle f(r) \rangle = V^{-1} \int_V f'(r, r') dr', \tag{4}$$

where r is the vector-radius of the point in the volume V and r' is the local system of coordinates with r as the origin.

The structure of MHM is characterized at each of the following scales [12]:

i) Microscale, l_0 (corresponds to the dimensions of microheterogeneities like crystals, disperse particles, etc.);

ii) Miniscale, l (corresponds to the dimensions of BRE);

iii) Macroscale, L (corresponds to specimen's dimensions).

The necessary and sufficient condition of the validity of the concept of 'effective properties', defined by the following inequality,

$$l_0 \ll l \gg L, \tag{5}$$

permits one to determine the effective conductivity λ with the aid of Eqs. (2) and (3) from the following system:

$$\lambda = \lambda_1 \varphi A_1 + \lambda_2 (1 - \varphi) A_2, \tag{6.a}$$

$$\varphi A_1 + (1 - \varphi) A_2 = 1, \tag{6.b}$$

where A_i are defined by

$$\langle \nabla t_i \rangle = A_i \langle \nabla t \rangle, \tag{7}$$

$$\langle \nabla t_i \rangle = V_i^{-1} \int\!\!\int\!\!\int_{V_i} \nabla t(r) \, dV, \ i = 1,2. \tag{8}$$

V_i is the volume occupied by i-th component and φ is the volume concentration of component 1.

A similar procedure may be applied to determine ρ, i.e.

$$\rho = \rho_1 \varphi B_1 + \rho_2 (1 - \varphi) B_2, \tag{9.a}$$

$$\varphi B_1 + (1 - \varphi) B_2 = 1, \tag{9.b}$$

where B_is are defined as

$$\langle \mathbf{q}_i \rangle = B_i \langle \mathbf{q} \rangle, \ \langle \mathbf{q}_i \rangle = V_i^{-1} \int\!\!\int\!\!\int_{V_i} q(r) \, dV, \ i = 1,2. \tag{10}$$

Equations (6) and (9) cannot be used for a straightforward calculation of λ and ρ, as the number of unknowns (λ, A_1 and A_2 in the former case, ρ, B_1 and B_2 in the latter) exceeds the number of equations available (two in each case). Thus additional information on the structure of MHM becomes necessary.

Consider the simplest structure, the stratified MHM. When the heat flux q is directed along the strata, one may write

$$\langle \nabla t_1 \rangle = \langle \nabla t_2 \rangle = \langle \nabla t \rangle, \tag{11}$$

and, therefore, since $A_1 = A_2 = 1$, one obtains from (6)

$$\lambda_\| = \lambda_1 \varphi + \lambda_2 (1 - \varphi), \tag{12}$$

where $\lambda_\|$ is the conductivity along the strata.

When the heat flux q is directed normal to the strata (i.e., $\langle \mathbf{q}_1 \rangle = \langle \mathbf{q} \rangle$, hence, $B_1 = B_2 = 1$), from Eq. (9) one obtains $\rho_\perp = \rho_1 \varphi + \rho_2 (1 - \varphi)$; finally, the conductivity in the direction normal to strata is derived as

$$\lambda_\perp = [\varphi/\lambda_1 + (1 - \varphi)/\lambda_2]^{-1}]. \tag{13}$$

Combining Eqs. (12) and (13), one obtains

$$\lambda_\| - \lambda_\perp = \frac{(\lambda_1 - \lambda_2)^2 \varphi (1 - \varphi)}{\lambda_1 \varphi + \lambda_2 (1 - \varphi)}. \tag{14}$$

As, generally, $\lambda_\perp \leq \lambda \leq \lambda_\|$, one may write

$$\lambda = \langle \lambda \rangle - K \frac{(\lambda_1 - \lambda_2)^2 \varphi (1 - \varphi)}{\lambda_1 \varphi + \lambda_2 (1 - \varphi)}. \tag{15}$$

where $0 \leq K \leq 1$ is the structure-dependent coefficient.

Thus structural characterization of MHM by corresponding A_is ad B_is ought to be the first step in any theoretical analysis of its effective conductivity.

2.2 Method of Direct Calculations

In this approach, the unknown parameters, A_is and/or B_is, are calculated directly from differential equations in partial derivatives for an appropriate model. In the simplest case of a dilute suspension of spherical inclusions of component 1 in a continuous matrix of component 2, the temperature field around any particle is presumed to be unperturbed by its neighbors (i.e., the external field gradient is equal to the mean gradient in MHM); hence, one may write

$$A_1 = 3\lambda_2/(\lambda_1 + 2\lambda_2). \tag{16}$$

Substitution of Eq. (16) into Eq. (6) yields [12]

$$\lambda = \lambda_2 + \frac{3\lambda_2 \varphi (\lambda_1 - \lambda_2)}{\lambda_1 + 2\lambda_2}. \tag{17}$$

Assuming, further, the equality of the mean gradients within the external field and the matrix, means replacement of Eq. (16) by Eq. (18),

$$A_1 = \frac{3\lambda_2}{\lambda_1 + 2\lambda_2} \left[\frac{3\lambda_2}{\lambda_1 + 2\lambda_2} \varphi + (1 - \varphi) \right]^{-1} = \frac{3\lambda_2}{3\lambda_2 \varphi + (\lambda_1 + 2\lambda_2)(1 - \varphi)}. \tag{18}$$

Finally, substituting Eq. (18) into Eq. (6), Maxwell's equation, Eq. (19) is obtained [13]

$$\lambda = \lambda_2 + \frac{\lambda_1 + 2\lambda_2 - 2\varphi(\lambda_2 - \lambda_1)}{\lambda_1 + 2\lambda_2 + \varphi(\lambda_2 - \lambda_1)}. \tag{19}$$

Lorenz [14] and Lorentz [15] examined the case of an inclusion separated from the continuous medium by an interlayer with intermediate conductivity to derive Eq. (20),

$$\frac{\lambda_2 - \lambda}{\lambda_2 + 2\lambda} = \varphi \frac{\lambda_2 - \lambda_1}{\lambda_2 + 2\lambda_1}. \tag{20}$$

Recently, Malyshev and Malyshev [16] derived Eq. (21) for the effective conductivity of MHM modelled by spherical inclusions located in the regular cubical lattice,

$$\lambda = \lambda_2 \left\{ 1 + 3\varphi \left[\frac{\lambda_1 + 2\lambda_2}{\lambda_1 - \lambda_2} - \varphi + \frac{A}{B} + C \right]^{-1} \right\} \tag{21}$$

where

$$A = 1.3091 \frac{\lambda_2 - \lambda_1}{\lambda_1 + 4\lambda_2/3} \varphi^{10/3} \left(1 - 0.1173 \frac{\lambda_2 - \lambda_1}{\lambda_1 + 6\lambda_2/5} \varphi^{11/3} \right)^2;$$

$$B = 1 + 0.4054 \frac{\lambda_2 - \lambda_1}{\lambda_1 + 4\lambda_2/3} \varphi^{7/3} - 6.6568 \frac{(\lambda_2 - \lambda_1)^2 \varphi^6}{(\lambda_1 + 4\lambda_2/3)(\lambda_1 + 5\lambda_2/6)},$$

$$C = 0.0723 \frac{\lambda_2 - \lambda_1}{\lambda_1 + 6\lambda_2/5} \varphi^{14/3} + 0.15256 \frac{\lambda_2 - \lambda_1}{\lambda_1 + 8\lambda_2/7} \varphi^6.$$

Truncation of Eq. (21) after the first two terms yields Eq. (17); Rayleigh's Eq. (22) [17] is recovered by truncation after the first four terms,

$$\lambda = \lambda_1 \left(1 + \frac{3\varphi}{\frac{\lambda_1 + 2\lambda_2}{\lambda_1 - \lambda_2} - \varphi - 1.3091 \frac{\lambda_1 - \lambda_2}{\lambda_1 + 4\lambda_2/3} \varphi^{10/3}} \right); \quad (22)$$

finally, Eq. (23) of Meredith and Tobias [18] is obtained by truncation after first six terms,

$$\lambda = \lambda_2 \left\{ 1 + 3\varphi \left[\frac{\lambda_1 + 2\lambda_2}{\lambda_1 - \lambda_2} - \varphi - \frac{1.3091 \frac{\lambda_1 - \lambda_2}{\lambda_1 + 4\lambda_2/3} \varphi^{10/3}}{1 - 0.4054 \frac{\lambda_1 - \lambda_2}{\lambda_1 + 4\lambda_2/3} \varphi} \right.\right.$$

$$\left.\left. - 0.0723 \frac{\lambda_1 - \lambda_2}{\lambda_1 + 6\lambda_2/5} \varphi^{14/3} \right]^{-1} \right\}. \quad (23)$$

A common feature of all cited approaches is the neglect of contact effects at the interface between components (i.e., ideal thermal contact is assumed).

2.3 Self-Consistent Field Approach

The basic model is an isolated spherical inclusion embedded into an infinite medium with the effective properties being sought. Making use of arguments similar to those used in derivation of Eq. (16), one may write

$$A_1 = 3\lambda/(\lambda_1 + 2\lambda_2), \quad (24)$$

to obtain, after substitution of Eq. (24) into Eq. (6) [19–21],

$$\lambda = \lambda_1 \{ [(3\varphi - 1) + (2 - 3\varphi)a]/4 + ([(3\varphi - 1) + (2 - 3\varphi)a]^2/16 + a/2)^{1/2} \}, \quad (25)$$

where $a = \lambda_2/\lambda_1$.

Extension of this approach to the case of inclusions of ellipsoidal shape yielded [22, 23]

$$\lambda = \lambda_2 + \frac{\varphi}{3(1 - \varphi)} \cdot \frac{\lambda_1 - \lambda_2}{1 + (b/2)(\lambda_1/\lambda_2 - 1)}, \quad (26)$$

where b is the numerical shape factor equal to zero, 2/3 and unity for sheets, spheres and rods, respectively. In the case of disc-like and rod-like shapes (i.e., compressed and stretched spheroids), this factor is defined by Eqs. (27) and (28):

$$b = \frac{\psi - 0.5 \sin 2\psi}{\sin 2\psi} \cos \psi, \qquad (27)$$

$$b = \frac{1}{\sin^2 \psi} - \frac{\cos^2 \psi}{\sin^3 \psi} \ln \frac{1 + \sin \psi}{1 - \sin \psi}, \qquad (28)$$

where $\psi = \arccos(d/l)$, d is the thickness and l is either the disc diameter, or the rod length.

This approach fails at $a < 10^{-2}$ [24, 25]; moreover, $\lambda < 0$ is predicted at $a = 0$ and $\varphi < 0.3$.

2.4 Method of Integration

The incremental change of the effective conductivity of MHM on addition of a small amount of inclusions may be defined as [cf. Eq. (16)]

$$\Delta\lambda = \lambda - \lambda_2 = \frac{3(\lambda_1 - \lambda_2)\lambda_2}{\lambda_1 + 2\lambda_2} \Delta\varphi, \qquad (29)$$

where $\Delta\varphi = \Delta V_1/V$, V_1 is the volume occupied by component 1 and ΔV_1 is the amount of component 1 added. Hence, the volume content of the latter will change from initial φ to φ^*, i.e.,

$$\varphi^* = (V_1 + \Delta V_1)/(V_1 + V_2 + \Delta V_1) = (\varphi + \Delta\varphi)/(1 + \Delta\varphi),$$

and therefore the incremental change of the volume content is

$$\Delta\varphi = \varphi^* - \varphi = (1 - \varphi)\Delta\varphi \Rightarrow d\varphi = (1 - \varphi)d\varphi. \qquad (30)$$

Substitution of Eq. (30) into Eq. (29) yields

$$d\lambda = \frac{3(\lambda_1 - \lambda_2)\lambda_2}{\lambda_1 + 2\lambda_2} \cdot \frac{d\varphi}{1 - \varphi} \qquad (31)$$

which, after substitution of λ for λ_2, transforms into

$$d\lambda = \frac{3(\lambda_1 - \lambda)\lambda}{\lambda_1 + 2\lambda} \cdot \frac{d\varphi}{1 - \varphi}. \qquad (32)$$

Integration of Eq. (32), assuming $\lambda_{\varphi=0} = \lambda_2$, finally gives [19]

$$\frac{\lambda - \lambda_1}{\lambda_2 - \lambda_1}(\lambda_2/\lambda)^{1/3} = 1 - \varphi. \qquad (33)$$

Equation (33) is intended to apply only for the case of isolated inclusions.

2.5 Method of Averaging Over the Ensemble

This method combines the technique of averaging over the ensemble of permissible configurations with the self-consistent field (i.e., effective medium) approach [26–29].

Essentially, the effective properties of MHM are derived from the analysis of the pattern of physical fields around a selected spherical inclusion embedded into the medium with effective properties. However, in contrast to the usual 'effective medium model', the inclusion is assumed to be surrounded by a concentric shell with properties dependent on the distance to the surface of the sample inclusion. In this case, the mean volume content of the disperse component 1 in the immediate vicinity of the sample inclusion, $\varphi(\mathbf{r} - \mathbf{r}')$, is expected to differ from that in more remote regions, φ; moreover, sample inclusions are assumed to have no effect on the pattern of distribution of other inclusions outside the spherical shell of thickness R around the sample inclusion. Thus, the local concentration is defined as

$$\varphi(x) = \varphi\sigma(x/R); \ (x/R) \in (1,3),$$

$$\sigma(x/R) = \frac{27 - 56(x/R) + 30(x/R)^2 - (x/R)^4}{16(x/R)},$$

and $\sigma(x/R) = 1$ if $(x/R) > 3$.

The final result is

$$\lambda = \lambda_1 [7a(1 - \varphi) + 17 + 7\varphi]^{-1} \{a(1 + 11\varphi) + 5 - 11\varphi$$
$$+ [a(1 + 11\varphi) + 5 - 11\varphi]^2 + [7a(1 - \varphi) + 17 + 7\varphi]$$
$$\times [a(5 + 7\varphi) + 7(1 - \varphi)]^{1/2}\}. \tag{34}$$

As this result was obtained from first principles it may be recommended as a test for the validity of approaches by other models.

A closely related problem of the rate of heat transfer from a heated body embedded into the two-component material was solved (assuming that the undisturbed temperature gradient varies inversely with the square of the distance from the heated body) to yield [30]

$$\lambda/\lambda_2 = 1 + 3(\alpha - 1)[\varphi + f(\alpha)\varphi^2 + O(\varphi^3)]/[\alpha + 2 - (\alpha - 1)], \tag{35}$$

where $f(\alpha) = \sum_{p=6}^{\infty} [(B_p - 3A_p)/(p - 3)2^{p-3}]$ is the function decreasing from $f(\alpha) = \text{const} = 0.108$ (at $\log \alpha < -2$) to zero (at $\log \alpha = 0$) and increasing again up to $f(\alpha) = \text{constant} = 0.504$ (at $\log \alpha > 2$), B_p and A_p are the known functions of $\alpha = \lambda_2/\lambda_1$. It can be shown [30] that Eq. (35) reduces to Eq. (19) on neglect of the term in square brackets in the numerator.

2.6 Lichtenecker's Approach

Assuming similar functional dependences of both resistivity ρ and conductivity λ, i.e.

$$\lambda = \Phi(\lambda_1 \cdot \lambda_2 \cdot \varphi), \quad \rho = \Phi(\rho_1 \cdot \rho_2 \cdot \varphi),$$

provided the following conditions hold,

$$\Phi(1/\rho_1 \cdot 1/\rho_2 \cdot \varphi) = 1/\Phi(\rho_1 \cdot \rho_2 \cdot \varphi),$$

$$\Phi(\rho_1 \cdot \rho_2 \cdot 0) = \rho_2, \quad \Phi(\rho_1 \cdot \rho_2 \cdot 1) = \rho_1,$$

Lichtenecker [31] derived Eqs. (36):

$$\rho = \rho_1^{\varphi} \rho_2^{(1-\varphi)}, \tag{36.a}$$

$$\lambda = \lambda_1^{\varphi} \lambda_2^{(1-\varphi)}. \tag{36.b}$$

The simplicity of Eqs. (36) is attractive; however, the limits of their validity are unclear as soon as the vanishing effective conductivity of MHM, predicted for the case when the conductivity of either of the components is zero, contradicts the experimental data.

2.7 Herring's Approach

Using expansion of all fluctuating parameters into Fourier series inside volume V, i.e.

$$\lambda(r) = \langle \lambda \rangle + \sum_{e}{}' \lambda_i \exp(ier), \tag{37.a}$$

$$q(r) = \langle q \rangle + \sum_{k}{}' q_k \exp(ikr), \tag{37.b}$$

$$\nabla t(r) = \langle \nabla t \rangle + \sum_{m}{}' \nabla t_m \exp(imr), \tag{37.c}$$

and eliminating terms $k = 0, l = 0$ and $m = 0$ in summations Σ', Eq. (38) was derived [32, 33]:

$$\lambda = \langle \lambda \rangle \left[1 - (1/3)\frac{\langle(\lambda - \langle\lambda\rangle)^2\rangle}{\langle\lambda\rangle^2} + (1/3)^2 \frac{\langle(\lambda - \langle\lambda\rangle)^3\rangle}{\langle\lambda\rangle^3} - \cdots + \right.$$

$$\left. (-1/3)^{n-2} \frac{\langle(\lambda - \langle\lambda\rangle)^n\rangle}{\langle\lambda\rangle^n} + \cdots \right], (n > 2). \tag{38}$$

This method does not differentiate between different shapes (spheres, cylinders, etc.); moreover, Eq. (38) converges more slowly the more different the properties of both components.

2.8 Correlation Approximation. Method of Conditional Moments

The local values of the flux, temperature gradient and heat conductivity may be defined as follows:

$$q(r) = \langle q \rangle + q^0(r); \quad \nabla t(r) = \langle \nabla t \rangle + \nabla t^0(r);$$

$$\lambda(r) = \langle \lambda \rangle + \lambda^0(r); \quad t(r) = \langle t \rangle + t^0(r),$$

where the quantities with superscript "0" are the random functions of coordinates for which the following condition holds:

$$\langle q^0 \rangle = \langle \nabla t^0 \rangle = \langle \lambda^0 \rangle = \langle t^0 \rangle = 0.$$

The conservation equation for a stationary heat flux ($\operatorname{div} q(r) = 0$) may be written as (assuming $\operatorname{div}(\langle \lambda \rangle \langle \nabla t \rangle) = 0$):

$$\langle \lambda \rangle \operatorname{div} \nabla t^0 \operatorname{div}(\lambda^0 \nabla t) = 0, \tag{39}$$

which can be transformed into

$$\langle \lambda \rangle \operatorname{div} \nabla t^0(r) = -f(r), \quad f(r) = -\operatorname{div}(\lambda^0 \nabla t). \tag{40}$$

The latter, Eq. (40), may be solved using Green's function to yield linked chaining with respect to the local temperature $t^0(r)$; hence, one needs to split that chaining which is characteristic of non-linear systems.

Using this approach and neglecting higher-than-second moments of $\langle \lambda \nabla t \rangle$ (the so called 'correlation approximation'), the effective heat conductivity of isotropic, homogeneous MHM with isometric components may be expressed as [11]:

$$\lambda = \langle \lambda \rangle - \varphi(1 - \varphi)(\lambda_1 - \lambda_2)^2 / 3\langle \lambda \rangle. \tag{41}$$

Equation (41) may be regarded as a special case of Eq. (38) truncated after the first two terms; it is best suited to the case of small differences in properties of components and/or low concentrations ($\varphi, (1 - \varphi) < 0.2$). As a natural step to eliminate these limitations, the higher moments were considered to yield [11]

$$\lambda = \langle \lambda \rangle - \varphi(1 - \varphi)(\lambda_1 - \lambda_2)^2 / [3\langle \lambda \rangle - (1 - 2\varphi)(\lambda_1 - \lambda_2)]. \tag{42}$$

It turned out, however, that the agreement between the latter, Eq. (42), and the experimental data was worse the larger the difference between the properties of components.

2.9 Variational Approach

This method was applied to estimate the higher and the lower bounds of the effective conductivity of MHM [34–39]. In one approach, these estimates were derived from the principle of minimum entropy production, i.e. [37]

$$dS/d\tau = - \iiint\limits_{(V)} T^{-1}(\nabla t \cdot \mathbf{q}) \, dV \geq 0. \tag{43}$$

In the case of a stationary temperature field, Eq. (43) may be transformed into

$$\delta \iiint\limits_{(V)} \lambda [\nabla t(\mathbf{r})]^2 \, dV = 0, \tag{44}$$

to arrive at the upper bound, Eq. (45), after substitution of $\nabla t(\mathbf{r}) = \langle \nabla t \rangle + \langle \nabla t^0(\mathbf{r}) \rangle$ for the local $\nabla t(\mathbf{r})$ into Eq. (44), i.e.

$$\lambda \leq \langle \lambda(\mathbf{n} + \nabla t^0)^2 \rangle, \tag{45}$$

where $\mathbf{n}$ is the unit vector parallel to $\langle \nabla t \rangle$ and $t^0(\mathbf{r})$ is the arbitrary limited function.

On the other hand, recalling that $1/\lambda = \langle q^2/\lambda \rangle / \langle q \rangle$, and having rewritten Eq. (43) as

$$\delta \left(\iiint\limits_{\langle V \rangle} \lambda^{-1} \mathbf{q}^2(\mathbf{r}) \, dV \right) = 0, \tag{46}$$

the lower bound is derived,

$$\lambda \geq \lambda^{-1} \langle [\mathbf{n} + (\nabla \times \mathbf{A})]^2 \rangle^{-1}, \tag{47}$$

where $\mathbf{A}$ is the arbitrary limited vector. Setting $\nabla t^0 = 0$ and $\nabla \times \mathbf{A} = 0$, one finally obtains from Eqs. (45) and (47)

$$\langle \lambda^{-1} \rangle^{-1} \leq \lambda \leq \langle \lambda \rangle. \tag{48}$$

In another approach [34–36], introducing a reference body with the shape and dimensions identical to those of a real MHM and with the conductivity tensor λ^c, one may define the vector of polarization as

$$P_i = q_i - q_i^c, \quad q_i = -\lambda_{ij} \nabla t_j, \quad \mathbf{q}^c = -\lambda_{ij} \nabla t_j^c, \tag{49}$$

where q_i and ∇t_j are the heat flux and the temperature gradient in MHM respectively.

Denoting by one prime (′) the deviations from the mean of the properties in a reference body and by two primes (″) the central arbitrary quantities, i.e.,

$$\lambda'_{ij} = \lambda_{ij} - \lambda^c_{ij}, \quad P''_i = P_i - \langle P_i \rangle, \tag{50}$$

and making use of Hashin-Shtrikman's variational principles [34], one may define the functional,

$$U = -0.5 \iiint\limits_V (\lambda^c_{ij} \langle \nabla t_i \rangle \langle \nabla t_j \rangle - P_i t_{ij} P_j + P''_i \nabla_i t''_j + 2 P_i \langle \nabla t_i \rangle) \, dV, \tag{51}$$

for which the stationary state is ensured provided the following main conditions,

$$\nabla t_i = \tau_{ij} P_j, \quad \tau_{ij} \lambda_{jk} \equiv \delta_{ik}, \tag{52}$$

and additional conditions,

$$\nabla_i [\lambda_{ij}^c \nabla t_j''(\mathbf{r}) + P_i''] = 0, \quad t''(\mathbf{r})|_s = 0, \tag{53}$$

are satisfied.

The stationary values of the functional at Eq. (51) are the absolute minimum and the absolute maximum in the cases of $\lambda_{ij}^c - \lambda_{ij} \geq 0$ and $\lambda_{ij}^c - \lambda_{ij} < 0$ respectively. The final result for MHM composed of isotropic and homogeneous components is

$$\lambda = \left[\sum_\alpha \varphi_\alpha (\lambda_\alpha + 2\lambda^c)^{-1} \right]^{-1} - 2\lambda^c. \tag{54}$$

In the case when the unknown λ^c is estimated from the following relationships [38]

$$\lambda_+^c = \varphi \lambda_2 + (1-\varphi)\lambda_1 = \lambda_1 \lambda_2 \langle \lambda^{-1} \rangle,$$

$$\lambda_-^c = [\varphi/\lambda_2 + (1-\varphi)/\lambda_1]^{-1} = \lambda_1 \lambda_2 / \langle \lambda \rangle,$$

the interval between the lower and upper bounds may be defined as

$$\left[\sum_{\alpha=1}^N \varphi_\alpha (\lambda_k + 2\lambda_-^c)^{-1} \right]^{-1} - 2\lambda_-^c \leq \lambda \leq$$

$$\left[\sum_{\alpha=1}^N \varphi_k (\lambda_\alpha + 2\lambda_+^c)^{-1} \right]^{-1} - 2\lambda_+^c. \tag{55}$$

On substitution of $\lambda_+^c = \lambda_1$ and $\lambda_-^c = \lambda_2 (\lambda_2 > \lambda_1)$ into Eq. (55), one obtains Hashin-Shtrikman's equation, Eq. (56), for two-component MHM [34],

$$\langle \lambda \rangle - \frac{\varphi(1-\varphi)(\lambda_1 - \lambda_2)^2}{\varphi \lambda_2 + (1-\varphi)\lambda_1 + \lambda_1} \leq \lambda \leq \langle \lambda \rangle - \frac{\varphi(1-\varphi)(\lambda_1 - \lambda_2)^2}{\varphi \lambda_2 + (1-\varphi)\lambda_1 + \lambda_2}. \tag{56}$$

Equations (55) and (56) should be compared to Eq. (57) [40, 41],

$$\left[\langle 1/\lambda \rangle - \frac{2(1/\lambda_1 - 1/\lambda_2)\varphi(1-\varphi)}{\langle 1/\lambda \rangle_x + 2\langle 1/\lambda \rangle_\varphi} \right]^{-1} \leq \lambda \leq \langle \lambda \rangle - \frac{(\lambda_1 - \lambda_2)^2 \varphi(1-\varphi)}{\langle \lambda \rangle_\varphi + 2\langle \lambda \rangle_x}, \tag{57}$$

(where $\langle \lambda \rangle_x = \lambda_1 x + \lambda_2(1-x)$, $\langle \lambda \rangle_\varphi = \lambda_1(1-\varphi) + \lambda_2 \varphi$ and $0 < x < 1$ is the geometrical factor), and to Eq. (58) [35, 42],

$$\lambda_1 + \varphi/[1/(\lambda_2 - \lambda_1) + \varphi/3\lambda_1] \leq \lambda \leq \lambda_2 + (1-\varphi)/$$
$$\times [1/(\lambda_2 - \lambda_1) + (1-\varphi)/3\lambda_2], \tag{58}$$

which was derived for the case of a single spherical inclusion with either higher (upper bound) or lower (lower bound) conductivity respectively, compared to the continuous medium.

It can be shown that the gap between upper and lower bounds predicted by Eqs. (55)–(58) increases and becomes comparable to the magnitude of λ_1 with $\lambda_1/\lambda_2 \Rightarrow 0$. One should also recognize the crucial influence of the structural model (in fact, method of the closure of equations based on the Green's function) chosen in each of the above variational approaches. If the Green's function may be defined for only the simplest shapes like spheres and cylinders, the theoretical predictions are also limited to MHM with heterogeneities of this kind. Thus it seems pertinent to emphasize once again that the theoretical predictions of the effective properties of MHM depend more on the structural model chosen, rather that on details of the mathematical formalism.

2.10 Method of Functional Inequalities

The starting point are Eqs. (2) and (3), which may be rewritten as [43]

$$\lambda \langle \nabla t(\mathbf{r}) \rangle^2 = \langle \lambda(\mathbf{r}) \nabla t^2(\mathbf{r}) \rangle, \qquad (59.a)$$

$$\rho \langle \mathbf{q}(\mathbf{r}) \rangle^2 = \langle \rho(\mathbf{r}) \mathbf{q}^2(\mathbf{r}) \rangle. \qquad (59.b)$$

Next, the problem of effective conductivity will be solved in the following steps.

i) Assuming the properties of Chebyshev's inequalities for the reciprocally ordered functions $f(\mathbf{r})$ and $g(\mathbf{r})$, i.e. [43]

$$[f(\mathbf{r}_1) - f(\mathbf{r}_2)][g(\mathbf{r}_1) - g(\mathbf{r}_2)] \leq 0,$$

$$\langle f(\mathbf{r}) g(\mathbf{r}) \rangle \leq \langle f(\mathbf{r}) \rangle \langle g(\mathbf{r}) \rangle,$$

also apply to functions $\lambda(\mathbf{r})$, $\nabla t^2(\mathbf{r})$, $\rho(\mathbf{r})$ and $\mathbf{q}^2(\mathbf{r})$, one may write

$$\langle \lambda \rangle \langle \nabla t^2 \rangle \geq \langle \lambda(\mathbf{r}) \nabla t^2(\mathbf{r}) \rangle = \lambda \langle \nabla t \rangle^2, \qquad (60.a)$$

$$\langle \rho \rangle \langle \mathbf{q}^2 \rangle \geq \langle \rho(\mathbf{r}) \mathbf{q}^2(\mathbf{r}) \rangle = \rho \langle \mathbf{q} \rangle^2, \qquad (60.b)$$

(where $1/\lambda = \rho$) to obtain

$$\langle 1/\lambda \rangle^{-1} \leq \lambda \leq \langle \lambda \rangle. \qquad (61)$$

(ii) Assuming, further, the applicability of Jensen's inequalities for a convex function $f(x)$ [45],

$$f(\langle x \rangle) < \langle f(x) \rangle,$$

$$1/\langle f(x)^{-1} \rangle \leq \exp \langle \ln f(x) \rangle \leq \langle f(x) \rangle,$$

(where $1/\langle f(x)^{-1} \rangle$ and $\exp \langle \ln f(x) \rangle$ are harmonic and geometrical averages respectively) to the effective conductivity of MHM (at least at sufficient distance from the phase transition point) the following inequality should hold:

$$1/\langle 1/\lambda \rangle \leq \exp \langle \ln \lambda \rangle \leq \langle \lambda \rangle. \qquad (62)$$

Comparing Eqs. (61) and (62), one may derive, approximately,

$$\lambda \cong \exp\langle \ln \lambda \rangle \cong \lambda_1^{\varphi_1} \lambda_2^{\varphi_2} \cdots \lambda_n^{\varphi_n}, \tag{63}$$

which is essentially Lichtenecker's Equation (36.b).

iii) Assuming the applicability of the reverse Gelder's inequality for functions $f(\mathbf{r})$ and $g(\mathbf{r})$ [46],

$$\langle f(\mathbf{r})g(\mathbf{r})\rangle > [\langle f^p(\mathbf{r})\rangle]^{1/p}[\langle g^l(\mathbf{r})\rangle]^{1/l},$$

(where $0 < p < 1$ or $p < 0$, p and l are the conjugated numbers, i.e., $1/p + 1/l = 1$) to functions $\lambda(\mathbf{r})$, $\nabla t^2(\mathbf{r})$, $\rho(\mathbf{r})$ and $q^2(\mathbf{r})$, one may write

$$[\langle \lambda(\mathbf{r})\nabla t^2(\mathbf{r})\rangle] > [\langle \lambda^p(\mathbf{r})\rangle]^{1/p}[\nabla t^{2l}(\mathbf{r})\rangle]^{1/l}, \tag{64.a}$$

$$[\langle \rho(\mathbf{r})q^2(\mathbf{r})\rangle] > [\langle \rho^l(\mathbf{r})\rangle]^{1/l}[\langle q^{2p}(\mathbf{r})\rangle]^{1/p}. \tag{64.b}$$

Substitution of Eq. (49) into Eq. (64) yields

$$\lambda \geq (\langle \lambda^p \rangle^{1/p})C_1, \tag{65.a}$$

$$\lambda \geq (\langle \lambda^{-1} \rangle^{1/l})C_2, \tag{65.b}$$

where $C_1 = (\langle \nabla t^{2l}\rangle^{1/l})/(\langle \nabla t^2\rangle)$ and $C_2 = (\langle q^{2p}\rangle)^{1/p}/(\langle q^2\rangle)$. It can be shown that $C_1 < 1$ if $l < 1$, and $C_2 < 1$ if $p < 1$; moreover, having derived $\lambda \leq \langle \lambda \rangle$, $1/\lambda \leq \langle 1/\lambda \rangle$ from Eq. (61), and $\lambda \geq \langle \lambda^p \rangle^{1/p}$, $1/\lambda \geq \langle \lambda^{-1} \rangle^{1/l}$ from Eqs. (4.65), one may write

$$\langle \lambda \rangle \geq \langle \lambda^p \rangle^{1/p}, \tag{66.a}$$

$$\langle 1/\lambda \rangle \geq \langle \lambda^l \rangle^{1/l}, \tag{66.b}$$

provided the following conditions hold:

$$\langle \lambda^p \rangle^{1/p}|_{p=-1} = 1/\langle 1/\lambda \rangle, \langle \lambda^{-1} \rangle^{-1/l}|_{l=1} = \langle \lambda \rangle. \tag{67}$$

Thus, the limits of p may now be established as

$$-1 \leq p < 0 \quad (0 < l \leq 1/2), \tag{68.a}$$

$$0 < p \leq 1/2 \quad (-1 \leq l < 0). \tag{68.b}$$

Combining Eq. (66.a) and Eq. (66.b), one obtains, finally,

$$\langle \lambda^p \rangle^{1/p} \leq \lambda \leq \langle \lambda^{-1} \rangle^{-1/l}. \tag{69}$$

The gap between the upper and lower bounds of Eq. (69) at constant conductivity ratio and volume content of both components is not fixed but depends on p; moreover, the validity of Eq. (69) is restricted to a certain subensemble of the total variety of possible values, $1/\langle 1/\lambda \rangle \leq \lambda \leq \langle \lambda \rangle$. This latter aspect will be analyzed below for the following special cases.

a) $p = -1, l = 1/2$,

$$1/\langle 1/\lambda \rangle < \lambda < 1/\langle \lambda^{1/2} \rangle^2. \tag{69.a}$$

b) $p = -1/2, l = 1/3$,
$$1/\langle \lambda^{1/2} \rangle^2 < \lambda < 1/\langle 1/\lambda^{1/3} \rangle^3. \tag{69.b}$$

c) $p = 1/3, l = -1/2$,
$$\langle \lambda^{1/3} \rangle^3 < \lambda < \langle \lambda^{1/2} \rangle^2. \tag{69.c}$$

d) $p = 1/2, l = -1$,
$$\langle \lambda^{1/2} \rangle^2 < \lambda < \langle \lambda \rangle. \tag{69.d}$$

All the above inequalities conform to Eq. (61); however, is seems obvious that different values of p and l refer to different MHM structures. In fact, Eq. (69.a) is expected to apply in the case of highly conductive inclusions embedded into a poorly conductive medium ($\lambda_1 \gg \lambda_2$), while Eq. (69.d) would be better suited in the reverse case; the intermediate situations would be described by Eqs. (69.b) and (69.c).

The relevance of arbitrary numbers p and l to the structural characteristics of random MHM may be assessed from Eq. (70),

$$p = 1/(1 + \delta), l = -1/\delta, \tag{70}$$

where $\delta = \langle \delta \rangle/d$, $\langle \delta \rangle = 1.19 d/\varphi^{1/3}$ is the mean distance between highly conducting inclusions and d is the mean linear dimension of the latter [39]. Substitution of Eq. (70) into Eq. (69) yields

$$\langle \lambda^{1/(1+\delta)} \rangle^{1+\delta} \le \lambda \le \langle \lambda^{1/\delta} \rangle^\delta, \tag{71}$$

which can be easily transformed into Eq. (72) for a two-component MHM;

$$[\varphi + a^{1/(1+\delta)}(1 - \varphi)]^{1+\delta} \le \lambda/\lambda \le [\varphi + a^{1/\delta}(1 - \varphi)]^\delta. \tag{72}$$

Equation (72) predicts a much narrower gap between upper and lower bounds of the effective conductivity of two-component MHM compared to other approaches (e.g., Eqs. (56)–(58) and therefore provides better agreement with the experimental data available for certain systems (Fig. 1). However, special analysis proved that the quantitative applicability of Eq. (72) was limited both by φ and by a (cases with $\varphi > 0.2$ and $a \ge 10^{-4}$ may be recommended).

2.11 Method of Integral Sectioning

Estimates of the upper and lower bounds within the framework of this method were obtained from general physical arguments rather than from rigorous mathematical treatment [8, 17, 32, 47, 48]. According to Eq. (43), the additional constraint, i.e. div $\mathbf{q}(\mathbf{r}) = 0$, ensures the stationary state of the integral (i.e., energy dissipation during the heat flux flow),

$$I = (2V)^{-1} \iiint_{(V)} \mathbf{q}(\mathbf{r}) \nabla t(\mathbf{r}) dx_1 dx_2 dx_3.$$

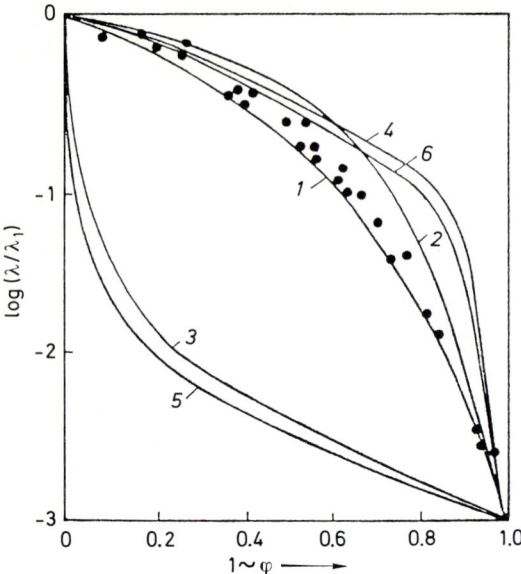

Fig. 1. Composition dependence of reduced electrical conductivity of NH_3–Li solution (assuming $a = 1.2 \times 10^{-3}$) and theoretical calculations by Eq. (72) (curves 1, 2), Eq. (56) (curves 3, 4) and Eq. (58) (curves 5, 6)

Moreover, the minimum value (i.e., $\delta I = 0$) corresponds to the series of legitimate function $[\mathbf{q}(\mathbf{r}), \nabla t(\mathbf{r})]$ which satisfy the conditions

$$\text{div}[\lambda(\mathbf{r})\nabla t(\mathbf{r})] = 0$$

$$\mathbf{q}(\mathbf{r}) = \lambda(\mathbf{r})\nabla t(\mathbf{r}).$$

Therefore, any other choice of pair functions, $[\mathbf{q}'(\mathbf{r}), \nabla t(\mathbf{r})]$ or $[\mathbf{q}(\mathbf{r}), \nabla t'(\mathbf{r})]$, from the series of legitimate functions which satisfy the same boundary conditions as do the true functions, $[\mathbf{q}(\mathbf{r}), \nabla t(\mathbf{r})]$, would yield the solution I' such that $I' \geq I$, where

$$I' = \langle \mathbf{q}'\nabla t \rangle, \langle \mathbf{q}' \rangle = \lambda'\langle \nabla t \rangle,$$

or

$$I' = \langle \mathbf{q}\nabla t' \rangle, \langle \nabla t' \rangle = \rho'\langle \mathbf{q} \rangle.$$

Here λ' and ρ' are the effective properties of fictive (or reference) bodies characterized by pair functions $[\mathbf{q}'(\mathbf{r}), \nabla t(\mathbf{r})]$ or $[\mathbf{q}(\mathbf{r}), \nabla t'(\mathbf{r})]$, respectively.

The following useful relationship was also proved to hold for quasi-homogeneous bodies [49]:

$$\langle \mathbf{q}(\mathbf{r})\nabla t(\mathbf{r}) \rangle = \langle \mathbf{q} \rangle \langle \nabla t \rangle.$$

Two methods for selecting the test functions $\mathbf{q}'(\mathbf{r})$ and $\nabla t'(\mathbf{r})$ to estimate upper and lower bounds of the effective conductivity of MHM are available. For convenience, operators of averaging by coordinates will be introduced,

$$\{f(r)\}_L = L^{-1} \int_0^L f(r) dx_k,$$

$$\{f(r)\}_S = S^{-1} \iint_S f(r) dx_i dx_j,$$

assuming validity of the following equalities:

$$\{\{f(r)\}_S\}_L = \{\{f(r)\}_L\}_S = \langle f(r) \rangle.$$

The two-bounds estimates by the integral sectioning approach are based essentially on two methods of conditional sectioning of the representative volume V. In the first case (i) the latter (Fig. 2a) is sectioned in the direction selected along the external field (say, along the axis Ox_3) into the prisms of $dx_1 dx_2$ for the basal area and L for the height (Fig. 2b), whereas in the second case (ii) the volume V is sectioned into layers of dx_3 for the thickness and $L \times L$ for the basal area (Fig. 2c).

i) The heat flux is expressed as

$$\langle q \rangle = \{\{\lambda(r) \nabla t(r)\}_S\}_L, \tag{72}$$

and the test functions $\{q'(r), \nabla t(r)\}$ are chosen such that the following condition will hold:

$$\{q'(r)\}_S = \langle q' \rangle. \tag{73}$$

Given Eq. (73), the mean heat flux through the specimen's cross-section [i.e., the term $\{...\}_S$ in Eq. (72)] may be defined as

$$\{q(r)\}_S \{\lambda(r) \nabla t(r)\}_S = [\lambda]_S [\nabla t(r)]_S, \tag{74}$$

where $[\lambda]_S = \lambda_1 \langle S \rangle_1 (x_3) + \lambda_2 \langle S \rangle_2 (x_3)$ is the conductivity of the layer of thickness dx_3, $S = S_1(x_3) + S_2(x_3)$ is the cross-sectional area of the BRE in a normal

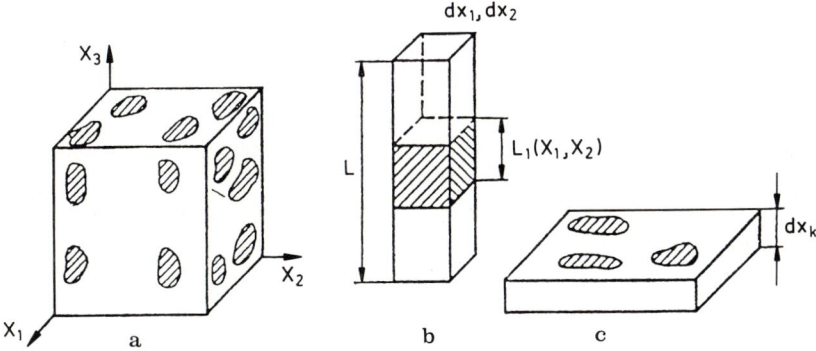

Fig. 2a–c. Structural models of MHM: **a** bulk representative element; **b** prism-shaped bulk differential element; **c** plate-shaped bulk differential element

direction to Ox_3 axis, $S_i(x_3)$ is that occupied by the i-th component (i = 1, 2), and $\langle S \rangle_i(x_3) = S_i(x_3)/S$.

Combining $\langle \nabla t \rangle = \{[\lambda]_S^{-1}\}_L \langle q' \rangle$ derived from Eq. (74) with $(\langle q \rangle \langle \nabla t \rangle) \leq (\langle q' \rangle \langle \nabla t \rangle)$, one obtains the upper bound,

$$\lambda \leq \{\{\lambda\}_S^{-1}\}_L^{-1}. \tag{75}$$

ii) The mean gradient is expressed as

$$\langle \nabla t \rangle = \{\{\nabla t(r)\}_L\}_S, \tag{76}$$

and the test functions $[\nabla t'(r), q(r)]$ are chosen to satisfy the condition

$$\{\nabla t'(r)\}_L = \langle \nabla t' \rangle. \tag{77}$$

Thus the term $\{...\}_L$ in Eq. (76) may be written as

$$\{\nabla t'(r)\}_L = \{\lambda^{-1}\}_L \{q(r)\}_L, \tag{78}$$

where $\{\lambda^{-1}\}_L = \langle L \rangle_1(x_1, x_2)/\lambda_1 + \langle L \rangle_2(x_1, x_2)/\lambda_2$ is the heat resistance of a prism of height L and of basal area $dx_1 dx_2$, $\langle L \rangle_i(x_1, x_2) = L_i(x_1, x_2)/L$, and $L_i(x_1, x_2)$ is the specimen's length along Ox_3 axis occupied by i-th component (i = 1, 2).

Using Eq. (78) to derive

$$\langle q \rangle = \{\{\lambda^{-1}\}_L^{-1}\}_S \langle \nabla t' \rangle, \tag{79}$$

one may write

$$I' = 0.5 \langle q \rangle \langle \nabla t' \rangle = 0.5 \{\{\lambda^{-1}\}_L^{-1}\}_S^{-1} \langle q \rangle^2. \tag{80}$$

Comparing this result with

$$I = 0.5 \langle q \rangle \langle \nabla t \rangle = 0.5 \lambda^{-1} \langle q \rangle^2 \tag{81}$$

and recalling that $I' \geq I$ (see above), one may write for the lower bound

$$\lambda \geq \{\{\lambda^{-1}\}_L^{-1}\}_S. \tag{82}$$

Combining Eq. (75) and Eq. (82), one finally defines upper and lower bounds of the effective heat conductivity of MHM, i.e.

$$\{\{\lambda^{-1}\}_L^{-1}\}_S \leq \lambda \leq \{\{\lambda\}_S^{-1}\}_L^{-1}. \tag{83}$$

Equation (83) is the mathematical formulation of the well known rule that the true effective conductivity of MHM is higher and lower than those estimated from "adiabatic" and "isothermal" sectioning respectively [8]. The physical meaning of the relevant assumptions (Eqs. (74) and (78) respectively), may be assessed from consideration of the flat circuit of two randomly connected resistances (Fig. 3a). The upper bound of the effective resistance (i.e., Eq. (74)) would then correspond to elimination of all transversal connections (Fig. 3b) since the finite resistances are replaced by infinite ones (circuit breakers). In its turn, the lower bound (i.e., Eq. (78)) would be obtained assuming zero resistance for all transversal connections (Fig. 3c).

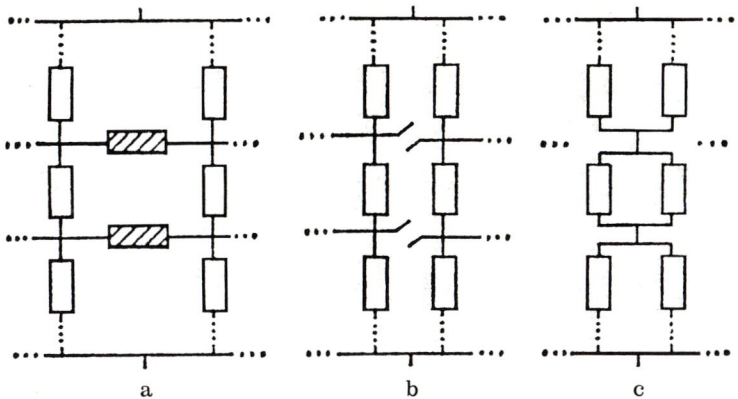

Fig. 3a–c. Randomly connected resistance circuit: **a** initial; **b** corresponding to Eq. (74); **c** corresponding to Eq. (78)

The general Eq. (83) was used to estimate the upper (λ_+) and the lower (λ_-) bounds of the effective conductivity of MHM with the following special unit cell geometries [8, 50].

i) Spherical inclusion in a cube:

$$\lambda_- = \lambda_2 + (\lambda^* - \lambda_2)\pi_2, \tag{84.a}$$

$$\lambda_+ = [(1 - \pi_1)/\lambda_2 + \pi_1/\lambda_c]^{-1}. \tag{84.b}$$

ii) Cubical inclusion in a cube:

$$\lambda_- = \lambda_2 \frac{\lambda_1 - (\lambda_1 - \lambda_2)(1 - \varphi^{2/3})\varphi^{1/3}}{\lambda_1 - \varphi^{1/3}(\lambda_1 - \lambda_2)}, \tag{85.a}$$

$$\lambda_+ = \lambda_2 \frac{\lambda_2 + (\lambda_1 - \lambda_2)\varphi^{1/3}}{\lambda_2 + (\lambda_1 - \lambda_2)\varphi^{2/3}(1 - \varphi^{1/3})}. \tag{85.b}$$

iii) Interpenetrating structures:

$$\lambda_- = \lambda_1[C^2 + a(1 - C)^2 + 2aC(1 - C)/(aC + 1 - C)], \tag{86.a}$$

$$\lambda_+ = \lambda_1 \bigg/ \left[\frac{1 - C}{C^2 + a(1 - C^2)} + \frac{C}{C(2 - C) + a(1 - C)^2}\right]. \tag{86.b}$$

Equations (86) may be compared with Eq. (87) derived using a somewhat modified method of unit cell sectioning [51]:

$$\lambda = \lambda_1 \left[\frac{C^2 + aC(1 - C)}{aC(1 - C) + 1 - C + C^2} + \frac{a[C(1 - C) + a(1 - c)^2]}{C(1 - C) + a(1 - C + C^2)}\right]. \tag{87}$$

In the above Eqs. (84)–(87), $\lambda^* = [2\lambda_2/(a - 1)\pi_1]\{1 - [1/(a - 1)\pi_1 \ln[(a - 1)\pi_1 + 1]\}$, $\lambda_c = \lambda_2(1 - 1/a)\pi_2/I_p(1)$, $\pi_1 = 2(3\varphi/4\pi)^{1/3}$ $\pi_2 = \pi^{1/3}(3\varphi/4)^{2/3}$, $p = 1 + a/(1 - a)\pi$, $a = \lambda_2/\lambda_1$, $I_b(z) = 0.5/b^{1/2} \ln[(z - b^{1/2})/(z + b^{1/2})](b > 0)$,

$I_b(z) = 1/|b|^{1/2} \arctan(z/|b|^{1/2})$ (b < 0), and the geometrical parameter C is obtained from the relationship, $3C^2 - 2C^3 = \varphi$.

3 Theoretical Aspects: Physical Approaches

3.1 Step-by-Step Averaging Approach

The concept of a sharp, "mathematical" (i.e., infinitely thin) interface between the components implicit in all "pragmatic" approaches considered so far had to be modified to allow for the existence of boundary interphase (BI) in two-component polymer systems. This "physical" aspect is, however, explicitly accounted for by the Step-by-Step Averaging (SSA) approach [52–54].

This approach is based essentially on the following postulates [8].

i) Transport properties of a binary system with disordered structure are identical to those of an appropriately chosen "mutually adequate" (as concerns isotropicity, mechanical stability, geometrical equivalence of components, etc.) one with ordered structure.

ii) The effective conductivities of an ordered, macroscopic binary system and of its microscopic unit cell are the same.

iii) Any multi-component system may be systematically reduced to a binary case.

Having defined the mutually adequate ordered system and its unit cell, one has to proceed with the SSA technique according to which any macroscopic property of MHM may be obtained as a final result of a step-by-step calculation of that property for each next level (or scale) of the whole spectrum of structural microheterogeneities involved. More precisely, one starts by definition of the lowest level of structural microheterogeneity (that is, the BI thickness, Δr) and calculates its effective property; a similar calculation is repeated at each next structural scale (i.e., BI of thickness Δr and a single particle of component 2 of the size 2r; isolated cluster IsC formed by several particles coated with BI, each of the latter assumed quasi-homogeneous; infinite cluster InC composed of quasi-homogeneous IsCs, etc.) until the macroscopic property sought for is obtained.

The practical application of the SSA technique requires incorporation of two basic models, the percolation model and the model of an equivalent element.

3.1.1 The Percolation Model

This model is intended to apply to a large variety of physical systems in which the geometrical phase transition (i.e., the reversible transition between a non-bonded and a bonded state) may occur [55–61]. The behavior of percolating systems is frequently studied numerically on the lattices which may be regarded as assemblies of bonds and sites. In this case, one would differentiate the problem of bonds from the problem of sites; namely, the former is concerned with monitoring the transition from a non-bonded to a bonded state with increasing concentration φ, whereas in the latter case such transition is studied on many lattice sites. The critical concentration φ_c corresponding to transition from isolated groups of particles (isolated clusters IsC) to an 'infinite' (i.e., spanning the whole lattice) cluster (InC) defines the "percolation threshold". In other words, $\varphi \geq \varphi_c$ is the domain of existence of InC, while only IsC are assumed to exist at $\varphi < \varphi_c$. The percolation thresholds for regular two- and three-dimensional lattices are shown in Table 1.

Consider the 'infinite' square lattice in which the probabilities that each pair of the nearest sites is either connected by a bond or not is φ and $1 - \varphi$, respectively. The characteristic dimension L_n of bonded regions (i.e., those with uninterrupted sequences of bonds between different sites) will be a rapidly rising function of φ. The InC spanning the whole lattice is created at φ_c, and further increase of the number of new bonds above φ_c, is accompanied by coalescence of IsCs to InC. The probability of coalescence to InC of IsCs will be higher the larger their size; therefore, in the region $\varphi > \varphi_c$, the IsCs of size L_c are assumed to be located in the discontinuities ("holes") of InC.

The mean-square gyration radius of a finite cluster is assumed to be

$$\langle L_n^2 \rangle = \sum_{i=1}^{n} |r_i - r_0|^2/n,$$

where $r_0 = \sum_{i=1}^{n} r_i/n$ is the apparent center of gravity, n is the number of sites in a finite cluster, r_i is the coordinate of i-th bond; the summation is carried out over all n bonds involved in IsC.

Table 1. Site- and bond-percolation thresholds in regular two- and three-dimensional lattices

Lattice type	Coordination number Z	Packing density φ	Percolation threshold		Critical number of bonds[a] $Z\varphi_c^b$	Critical volume fraction $\eta \phi_c^s$
			φ_c^b	φ_c^s		
traingular	3	0.61	0.6527	0.70	1.96	0.427
square	4	0.79	0.500	0.59	2.00	0.466
hexagonal	6	0.91	0.3473	0.50	2.08	0.455
tetrahedral (diamond)	4	0.34	0.39	0.43	1.56	0.143
cubic:						
simple	6	0.52	0.25	0.31	1.50	0.161
body-centered	8	0.68	0.18	0.24	1.44	0.163
face-centered	12	0.74	0.12	0.20	1.44	0.148
dense hexagonal	12	0.74	0.12	0.20	1.44	0.148

[a] For three-dimensional lattices, $Z\varphi_c^b \cong d/(d - 1)$.

The correlation function for a random array of mass points defined as

$$K(\mathbf{r}, \mathbf{r}') = \frac{1}{n}\sum_{i=1}^{n} g(\mathbf{r}'_i)g(\mathbf{r}_i + \mathbf{r}'_i) = \frac{\langle g(\mathbf{r})g(\mathbf{r} + \mathbf{r}')\rangle}{\langle g(\mathbf{r})\rangle}$$

may be considered as a measure of the mean density of a cluster at a distance $|\mathbf{r}'|$ from an arbitrary point r.

The correlation length χ (i.e., the measure of cluster connectivity on a percolating lattice) is defined as

$$\chi = \left\langle \left[\frac{1}{n}\sum_{i=1}^{n}(\mathbf{r}_i - \mathbf{r}_0)^2\right]^{1/2} \right\rangle.$$

Let S_n be the number of n-bonds clusters per one bond of an infinite lattice; then the sum $\sum nS_n$ will be the fraction of the bonds belonging to finite clusters, and the ratio,

$$P_n = \frac{nS_n}{\sum_n nS_n},$$

is the probability that an arbitrarily chosen bond belongs to the n-bonds cluster.

The mean number of bonds (i.e., the mean mass) in a finite cluster is

$$\langle M \rangle = \sum_{i=1}^{n} n^2 S_n / \sum nS_n.$$

Given these results, the correlation length for the case $\varphi < \varphi_c$ may be now expressed as

$$\chi = \left[\sum_{n=1}^{N} L_n^2 n^2 S_n \Big/ \sum_{n=1}^{N} n^2 S_n \right]^{1/2},$$

whereas χ goes to infinity on the approach to the percolation threshold (i.e., $|\varphi - \varphi_c| = |\Delta\varphi| \Rightarrow 0$) as

$$\chi = |\Delta\varphi|^{-\nu}, \tag{88}$$

where ν is the corresponding critical index.

In these conditions (i.e., when the system is in the vicinity of φ_c), the density of InC (i.e., the probability that an arbitrarily chosen bond belongs to InC),

$$P_\infty(\varphi) = n_k / \sum_{i=1}^{n} n_i,$$

also scales with $\Delta\varphi$ as

$$P_\infty(\varphi) \sim |\Delta\varphi|^\beta, \tag{89}$$

where β is the relevant critical index. Comparing Eq. (88) and Eq. (89), one obtains

$$P_\infty(\varphi) \sim \chi^{-\beta/\nu}. \tag{90}$$

It is pertinent to point out here that, in the scale interval between the lattice constant, l_0, and the correlation length, χ, all properties of InC are similar to those at the critical point. More precisely, in the interval, $l_0 \ll L \ll \chi$, the InC acquires the property of self-similarity (or scale invariance) which is a characteristic feature of 'fractals'.

The critical indices for percolating systems may be evaluated within the framework of two theoretical models, i.e., the bond-site model (i) [60] and the blob model (ii) [61].

i) The InC is visualized as a network with correlation length χ as a characteristic mesh size (Fig. 4a). The network junctions (sites) are connected by macrobonds with the contour length

$$L_b \sim |\Delta\varphi|^{-\xi}, \tag{91}$$

which may exceed χ. According to numerical estimates [55, 56], the critical index $\zeta = 1$ for any space dimensionality d, whereas $v = 1.33$ or 0.8–0.8 for $d = 2$ or 3, respectively. Comparing Eq. (88) and Eq. (91), one may conclude

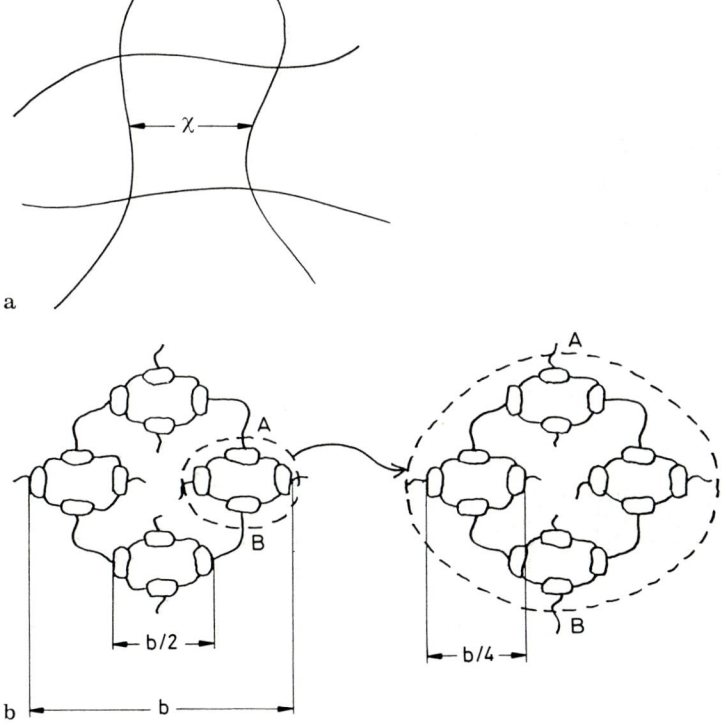

Fig. 4. a Bond-site model. b Blob model

that on the approach to the critical state χ is predicted to increase faster compared to L_b.

(ii) The InC is modelled as an array of blobs connected by macrobonds (Fig. 4b). The structure is assumed to duplicate itself at each next length scale, namely, each blob of the size b consists of several smaller blobs of the size b/2 connected by single macrobonds, etc.; thus at each structural scale the system is self-similar. It could be shown [61] that the mean cumulative length of non-overlapping bonds L_b in any portion of InC scales as $L_b \sim b^{1/\nu}$. Therefore, setting $b = \chi$, one obtains

$$L_b \sim \chi^{1/\nu} \sim |\Delta\varphi|^{-1}, \tag{92.a}$$

$$L_b/b \sim b^{1/(\nu-1)} \sim |\Delta\varphi|^{\nu-1}, \tag{92.b}$$

Thus, in contrast to case i) [cf. Eq. (91)], the relative bond length L_b/b is predicted to increase with scale length b.

The above data may now be used to analyze the conductivity Λ of the percolating system. Let the latter be sectioned into d-dimensional cubes (L > b) of linear dimensions $b \sim \chi$ and conductivity $\Lambda_b = \Lambda_1/b$ (where Λ_1 is the conductivity of a single bond). The number of cubes per unit length of a row is 1/b and the number of parallel rows is b^{1-d}; therefore the total conductivity will simply be $\Lambda = \Lambda_b b^{2-d} = \Lambda_1 b^{1-d}$. Setting $b = \chi$ one obtains

$$\Lambda = \Lambda_1 |\Delta\varphi|^{\nu(d-1)} = \Lambda_1 |\Delta\varphi|^t, \tag{93}$$

where $t = \nu(d - 1)$ is the critical conductivity index.

In one of the first attempts to account explicitly for percolation effects in conductivity of MHM [24, 25], three concentration intervals were considered for the case of highly conductive inclusions in a poorly conductive matrix ($\Lambda_2/\Lambda_1 \leq 5.10^{-4}$):

(i) $\varphi < \varphi_c \Rightarrow \Lambda = \Lambda_1(1 - 5\varphi)^{-1}$;
(ii) $\varphi_c \leq \varphi \leq 0.5 \Rightarrow \Lambda = 1.6\Lambda_1(\varphi - \varphi_c)^{1.6}$;
(iii) $\varphi > 0.5 \Rightarrow$ Eq. (25).

In the case of not-too-different conductivities of components ($3.10^{-2} \leq \Lambda_2/\Lambda_1 < 1$), the percolation effects were considered insignificant, and Eq. (25) for effective medium model was recommended throughout.

Another attempt to account explicitly for the percolation effects within the framework of the effective medium model yielded the following result [62]:

$$\varphi \frac{\lambda_1 - \lambda}{\lambda_1 + 2\lambda} + (1 - \varphi) \frac{\lambda_2 - \lambda}{\lambda_2 + 2\lambda} = 0. \tag{94}$$

Equation (94) was expected to apply in the case of not very different conductivities of components; moreover, it was derived assuming $\varphi_c = 0.3$ which differs significantly from the best theoretical estimates (cf. Table 1). In this respect, much more promising seems to be the approach based on the real space renormalization group (RSRG) theory [63] which provides for reasonable values of the percolation threshold (0.62 and 0.15 for two-dimensional, square tesselation and for three-dimensional, cubic tesselation, respectively) and is intended to apply for MHM whose components differ in their intrinsic conductivities by two and more orders of magnitude.

Apparently the same conclusion applies to the alternative approach which is conceptually similar to the RSRG method but is based on a slightly different tesselation technique [58, 59]. At small concentrations ($0 \leq \varphi \leq \varphi_c$), each IsC represented by a cubical inclusion with linear dimension $l_2 = \varphi^{1/3}$ is separated by distance $2(1 - l_2)$ from its neighbors (Fig. 5); the jump-like transition into InC by the mechanism of linking together the individual IsCs by single bonds of cross-sectional area $S = [(\varphi - \varphi_c)/(1 - \varphi_c)]^t$ is assumed at higher concentrations ($\varphi > \varphi_c$); in the intermediate interval, $\varphi \cong 1 - \varphi \cong 0.5$, each component creates its own InC; however, in the interval $\varphi > 0.5$ the InC of component 2 is assumed, first, to decrease in size, to degenerate, subsequently, into IsC at $\varphi = 1 - \varphi_c$ and, finally, to disappear at $\varphi \Rightarrow 1$. The relevant geometrical parameters of the percolation model discussed above are collected in Table 2.

The effective properties of the percolation model (Fig. 5) are derived by sectioning the BRE into four sections with faces parallel and/or perpendicular to the external fields. As an example, consider the case $\varphi < 0.5$ with geometrical parameters specified by Table 3. Let section I be filled with component 1; hence, the conductivity of this section is $\lambda_I = \lambda_1$. In sections II and III the components are connected in series; hence,

$$\lambda_{II} = \{1/\lambda\}_{l_2}^{-1} = [l_2/\lambda_1 + (1 - l_2)/\lambda_2]^{-1},$$

$$\lambda_{III} = \{1/\lambda\}_{l_1}^{-1} = [l_1/\lambda_1 + (1 - l_1)/\lambda_2]^{-1}.$$

Finally, $\lambda_{IY} = \lambda_2$ for section IY.

Thus, the effective conductivity of the percolation model will be obtained by averaging out the contributions from all four sections assuming normal orientation of the heat flux to their cross-sectional areas, i.e.,

$$\lambda = \lambda_1 S_1 + \lambda_{II} S_2 + \lambda_{III} S_3 + \lambda_2 S_4. \tag{95}$$

As already mentioned, Eq. (95) was derived for the case $\varphi \leq 0.5$; in the opposite case (i.e., $\varphi > 0.5$) one should change indices in Eq. (94) and in Table 2, i.e., $\lambda_1 \Leftrightarrow \lambda_2$ and $\varphi \Leftrightarrow 1 - \varphi$. The quality of the fit of Eq. (95) to the relevant experimental data may be assesssed from Fig. 6 [54].

Table 2. Geometrical parameters of the percolation model[a]

Range of validity	$\langle S_1 \rangle$	$\langle S_2 \rangle$	$\langle S_3 \rangle$	$\langle S_4 \rangle$	$\langle l_1 \rangle$	$\langle l_2 \rangle$
$0 \leq \varphi \leq \varphi_c$	0	$\varphi^{2/3}$	0	$1 - \varphi^{2/3}$	0	$\varphi^{1/3}$
$\varphi \leq \varphi \leq 0.5$	$\dfrac{\varphi - \varphi_c}{1 - \varphi_c^{1/3}}$	$\varphi_c^{2/3} - \langle S_1 \rangle$	$2\langle l_1 \rangle \langle l_2 \rangle$	$1 - \varphi_c^{2/3} - \langle S_3 \rangle$	$S_1^{1/2}$	$\varphi_c^{1/3}$

[a] $S_1 = \langle S \rangle + (\langle S_1 \rangle - \langle S \rangle)g(a)$; $\langle S \rangle = [(\varphi - \varphi_c)/(1 - \varphi_c)]^t$; $g(a) = 5.53a - 8.3a^2 + 3.23a^3 + 0.54a^4$

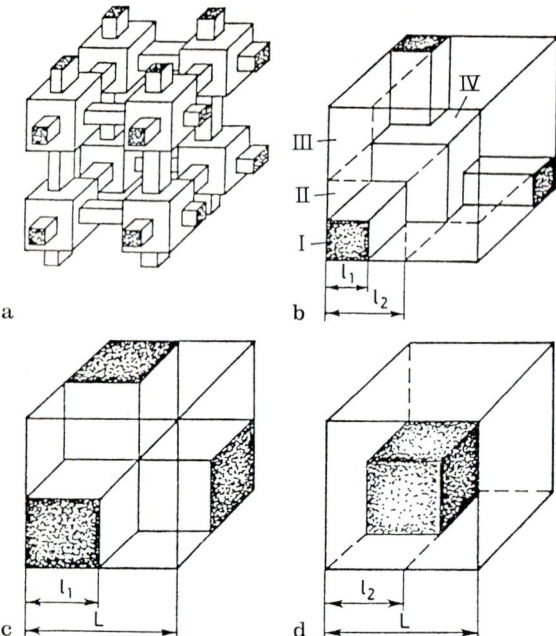

Fig. 5. a Percolation model. **b** Unit cells of BRE at $\varphi > \varphi_c$. **c** Unit cells of BRE at $\varphi \cong 0.5$. **d** Unit cells of BRE at $\varphi < \varphi_c$

Table 3. Geometrical parameters and volume fractions of components in different sections of BRE

Section no	Basal area	Renormalized volume contents	
		Component 1	Component 2
I	$\langle 1 \rangle^2$	1	0
II	$\langle l_2 \rangle^2 - \langle l_1 \rangle^2$	$\langle l_2 \rangle$	$1 - \langle l_2 \rangle$
III	$2\langle l_1 \rangle(1 - \langle l_2 \rangle)$	$\langle l_1 \rangle$	$1 - \langle l_1 \rangle$
IV	$1 - \langle l_2 \rangle^2 - 2\langle l_1 \rangle(1 - \langle l_2 \rangle)$	0	1

3.1.2 Voronoi Polyhedra. The Model of Equivalent Element

As already emphasized, any structural model of heterogeneous polymers claiming validity had to account explicitly for the smearing out of any sharp 'mathematical' interface into 'physical' BI of finite thickness Δr. Apparently, this requirement was met by the model of equivalent element (EE) [54, 64, 65] which was based on the following physical ideas (Fig. 7). The initially isolated particles of component 1 coated with BI of thickness Δr are assumed to coalesce, first, into IsC with a constant limiting volume content of component 1, φ^*, which may vary from about 0.3 for particles of very irregular shape to 0.6 for the dense random packing of spheres [56, 67]. The concentration φ' of such IsCs with

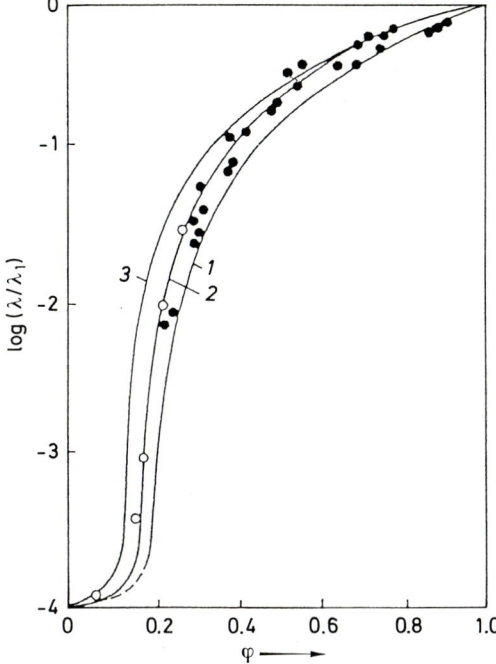

Fig. 6. Composition dependence of the relative heat conductivity of tungsten bronzes at 300 K. *Solid lines* were calculated by Eq. (11) with $\varphi_c = 0.25$, $t = 2$ (1), $\varphi_c = 0.15$, $t = 1.6$ (2) and $\varphi_c = 0.12$, $t = 0.6$ (3)

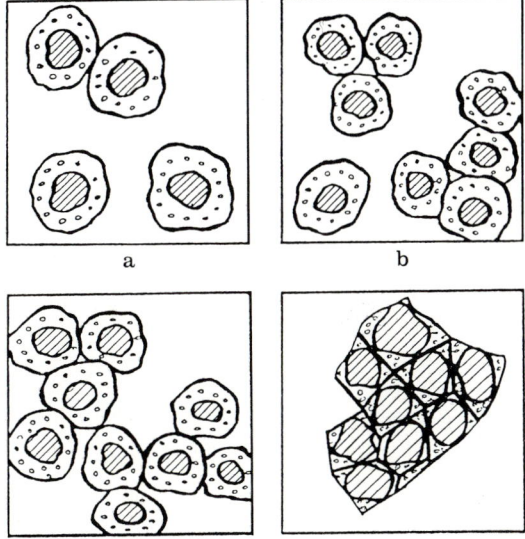

Fig. 7a–d. Schematics of MHM structure: **a** at $\varphi < \varphi_c$; **b** at $\varphi \cong \varphi_c$; **c** at $\varphi > \varphi_c$; **d** of tesselation of InC into Voronoi polyhedra

φ* = constant will increase with the nominal concentration φ until an InC is formed at φ' = φ_c. The BRE of the latter is represented by Voronoi polyhedra which are constructed by intersection of planes drawn normal to the vectors connecting the centers of the neighboring particles at their midpoints. In this fashion a system of Voronoi polyhedra is formed with their faces tangent to the points of BI contact. In absence of actual contact points in InC, the faces of polyhedra should be drawn such as to become eventually tangent to the points of contact after the appropriate smooth shift of particle centers (Fig. 7d). As a result, the InC is sectioned into different polyhedra with the number of faces dependent on the coordination number N_c of the corresponding particle. In other words, the BRE for this structure would be a polyhedron with the number of faces equal to the mean value of N_c which is defined as [52–54]

$$N_c = [(1 - \varphi^*) + 3 + [(1 - \varphi^*)^2 - 10(1 - \varphi^*) + 9]^{1/2}/2(1 - \varphi^*) \quad (96)$$

It becomes obvious that the limiting volume content of component 1, φ*, corresponds to perfect, void-free filling of the total space available by polyhedral BREs.

The effective properties of such BRE will be determined for a simplified case of a spherical particle embedded into a cube assuming the distance between the opposite faces of the latter smaller than the particle diameter 2r (Fig. 8). In this case, one should differentiate between the interparticle gap, Δl, and the BI thickness, Δr; in fact, there might be situations when either Δl = Δr (presumably in the case of strong interfacial interactions when all continuous phase available is transformed into BI), or Δr = 0 and Δl ≠ 0 (in the other extreme case of very weak interfacial interactions). The effective conductivity of BRE (Fig. 9) will be estimated as follows. Having applied the voltage to the opposite faces of the cube, the resulting current will comprise three contributions from the currents running through the particle (contact area S_1), through the matrix and the particle connected in series (cross-sectional area S_{12}) and through the matrix (cross-sectional area S_2), respectively.

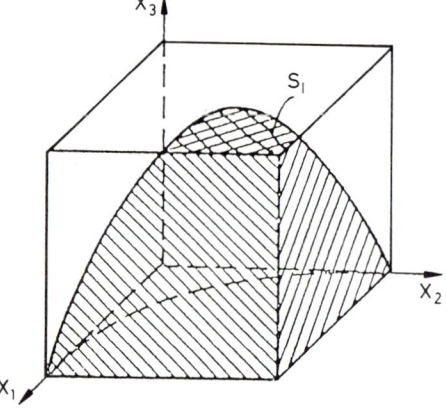

Fig. 8. Portion of a sphere embedded into a cube

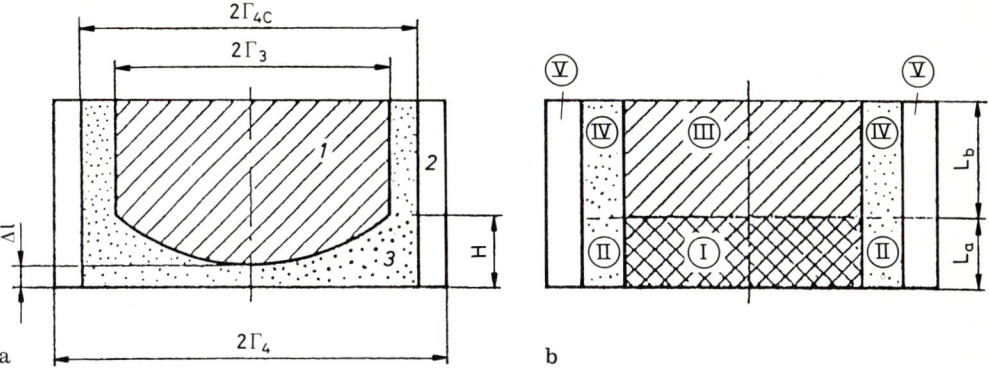

Fig. 9. a Model of equivalent element. **b** Scheme of its partitioning. 1-filler particle, 2-polymer, 3-boundary interphase

As soon as the three areas, S_1, S_{12} and S_2, specify the interparticle contacts in a polyhedron with any number of faces, it is convenient to model the contact region by a cylindrical body (i.e., the inner current channel) which is, in fact, the EE [8, 64, 65, 68, 69].

The EE appropriate for polymeric MHM is shown in Fig. 9. Section I comprises a spherical segment of component 1 inscribed into a cylinder and a gap of thickness Δl; thus, its height and basal area may be defined as $h_I = r + \Delta l - (r - r_3^2)^{1/2}$ and $S_I = \pi r_3^2$, respectively. Similar quantities for the remaining sections, as well as other pertinent geometrical parameters of EE are defined below.

$h_{II} = h_I$, $S_{II} = \pi(r_4'^2 - r_3^2)$;

$h_{III} = (r^2 - r_3^2)^{1/2}$, $S_{III} = S_I$;

$h_{IV} = h_{III}$, $S_{IV} = S_{II}$;

$h_Y = h_I + h_{II}$, $S_Y = \pi r_4^2$.

$h' = h/r = 1 + \Delta l' - (1 - y_3^2)^{1/2}$; $y_3 = r_3/r = 2(N_c - 1)^{1/2}/N_c$;

$y_4 = r_4/r = y_3/\varphi^{*1/3}$; $y_{4c} = r_{4c}/r = y_3/(1 - v_{BI})^{1/3}$; $\Delta l' = \Delta l/r$;

$v_{BI} = \varphi^*[(1 + \Delta r/r)^3 - 1]$.

The SSA technique will be now applied to evaluate the contributions to the effective conductivity of EE from the following structures:

a) section I $\Rightarrow \lambda_a$;
b) sections (I + II) $\Rightarrow \lambda_b$;
c) sections (III + IY) $\Rightarrow \lambda_c$;
d) sections (I − II) + (I − II) + (III + IY) $\Rightarrow \lambda_d$;
e) total EE [Y + (I + II) + (III + IY)] $\Rightarrow \lambda$.

The effective conductivity of section I will be determined by Eq. (97).

$$\lambda_a = S_1^{-1} \int\int_{(S_1)} \lambda(x_1, x_2) dx_1 dx_2 \tag{97}$$

which after transformations to polar coordinates and to those normalized by the mean radius of inclusions r will be replaced by Eq. (97.a),

$$\lambda_a = (\pi y_3)^{-1} \int_0^{2\pi} d\theta \int_0^{y_3} \lambda(\rho)\rho d\rho, \tag{97.a}$$

where $\lambda(\rho)$ is the conductivity of the differential cylindrical element (DCE) of section I which may be evaluated using Eq. (82) for the lower bound, i.e.,

$$\lambda(\rho) = \{[\lambda^{-1}]_{L_1(\rho)}^{1,BI}\}^{-1}, \tag{98}$$

where the superscripts 1.BI refer to components (component 1 and BI respectively) over which the averaging is performed, and the subscript $L_1(\rho)$ defines the geometry of the averaging procedure.

The expanded form of Eq. (98) is

$$\lambda(\rho) = \{L_1(\rho)/\lambda_1 + [1 - L_1'(\rho)]/\lambda_{BI}\}^{-1}, \tag{99}$$

where λ_{BI} is the conductivity of BI, $L_1'(\rho) = [(1 - y_3^2)^{1/2} - z_2]/(z_1 - z_2)$ is the length of DCE normalized by the total length of section I, $z_1 = 1 + \Delta l'$ and $z_2 = (1 - y_3)^{1/2}$.

Substitution of Eq. (99) into Eq. (97.a) yields

$$\lambda_a = B[\lambda_{BI}(z_1 - z_2)/y_3^2], \tag{100}$$

where

$$B = \frac{2}{(1 - a_{BI})^2} \left\{ [1 + \Delta l' - a_{BI}(1 - y_3^2)^{1/2}] \right.$$

$$\times \ln \frac{1 + \Delta l' - (1 - z_1^2)^{1/2} + a_{BI}X}{\Delta l' + a_{BI}[1 - (1 - y_3^2)^{1/2}]}$$

$$\left. - [1 - (1 - z_1^2)^{1/2}](1 - a_{BI}) \right\};$$

$X = (1 - z_1^2)^{1/2} - (1 - y_3^2)^{1/2}$; $a_{BI} = \lambda_{BI}/\lambda_1$.

Averaging the conductivity over the sections (b), (c) and (d) yields, respectively:

$$\lambda_b = \{\lambda\}_{\langle S_I \rangle}^{(a,BI)} = \lambda_a \langle S_I \rangle + \lambda_{BI}(1 - \langle S_I \rangle), \tag{101}$$

$$\lambda_c = \{\lambda\}_{\langle S_I \rangle}^{(1,BI)} = \lambda_1 \langle S_I \rangle + \lambda_{BI}(1 - \langle S_I \rangle), \tag{102}$$

$$\lambda_d = \{\lambda^{-1}\}_{L_a'}^{(b,c)} = [L_a'/\lambda_b + (1 - L_a')/\lambda_c]^{-1}, \tag{103}$$

where $\langle S_I \rangle = (r_3/r_{4c})^2$ is the normalized cross-sectional area of section I.

Finally, averaging over sections Y and (d) yields the effective conductivity of EE,

$$\lambda_{EE} = \{\lambda\}_{\langle S_{II}\rangle}^{(Y,\,d)} = \lambda_Y \langle S_{II}\rangle + \lambda_d (1 - \langle S_{II}\rangle), \tag{104}$$

where $\langle S_{II}\rangle = (r_{4c}/r_4)^2$ is the normalized cross-sectional area of the region b + c.

4 Tests of Model Predictions

4.1 Analytical Test

The influence of the structural parameters involved in the EE model on the heat conductivity of MHM will be assessed now from numerical calculations for the following representative cases (see below).

i) Fixed parameters: $\lambda_2 = 0.2$ W/m.K, $\varphi_c = 0.15$, $\varphi^* = 0.6$, $t = 1.8$, $\Delta l/r = 0.01$; variable parameters: λ_1, λ_{BI}/λ_2.

A common feature of all theoretical plots (Fig. 10) is the onset of an accelerated rise of λ at φ_c which becomes more prominent the higher the ratio λ_{BI}/λ_2, and the occurrence of another relatively weak discontinuity located symmetrically at $1 - \varphi_c$. Apparently the pattern of the composition dependence of λ is affected more by the ratio λ_{BI}/λ_2 than by λ_1/λ_2 (e.g., negative slope is invariably observed whatever the latter ratio if $\lambda_{BI}/\lambda_2 < 1$).

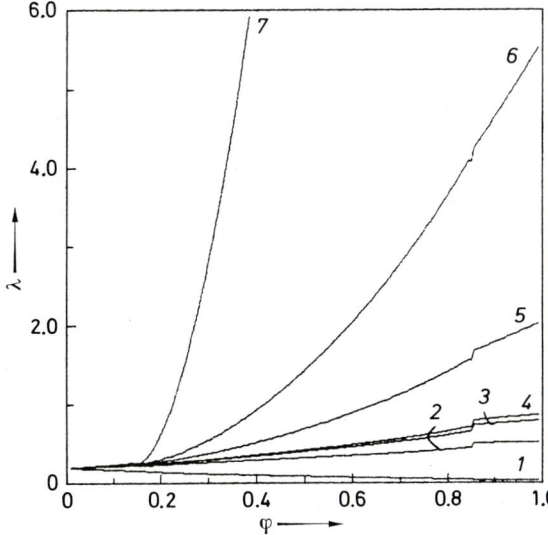

Fig. 10. Composition dependence of the heat conductivity of MHM for case i) calculated assuming $\lambda_1 = 2\text{-}200$ and $\lambda_{BI}/\lambda_2 = 0.1$ (*curve 1*); $\lambda_1 = 2, 20, 200$ and $\lambda_{BI}/\lambda_2 = 1$ (*curves 2, 3, 4*); $\lambda_1 = 2, 20, 200$ and $\lambda_{BI}/\lambda_2 = 10$ (*curves 5, 6, 7*)

ii) Fixed parameters: $\lambda_2 = 0.2$ W/m.K, $\lambda_1 = 2$ W/m.K, $\varphi^* = 0.6$, $\varphi_c = 0.15$, $t = 1.8$; variable parameters: λ_{BI}/λ_2, $\Delta l/r$.

The discontinuity at $1 - \varphi_c$ becomes sharper and the slope becomes higher the smaller the ratio $\Delta l/r$ (Fig. 11). As before, the larger the λ_{BI}/λ_2 ratio, the higher becomes the slope.

iii) Fixed parameters: $\lambda_2 = 0.2$ W/m.K, $\lambda_1 = 2$ W/m/K, $\lambda_{BI}/\lambda_2 = 1$, $\varphi_c = 0.15$, $\Delta l/r = 0.01$, $t = 1.8$; variable parameter: φ^*.

Both the slope and the jump at $1 - \varphi_c$ tend to increase the lower is φ^* (Fig. 12).

iv) Fixed parameters: $\lambda_2 = 0.2$ W/m/K, $\lambda_1 = 2$ W/m.K, $\lambda_{BI}/\lambda_2 = 1$, $\varphi^* = 0.6$, $\Delta l/r = 0.01$, $t = 1.8$; variable parameter: φ_c.

As might have been expected, the general pattern of the theoretical curves remained essentially unchanged, except the location of discontinuities at φ_c and at $1 - \varphi_c$ (Fig. 13).

Summarizing, the most crucial influence on the pattern of λ vs φ plots appears to be exerted by the ratios λ_{BI}/λ_2 and $\Delta l/r$. The point of concern is, however, the failure of most of the theoretical curves to extrapolate to λ_1 as φ approaches unity above $1 - \varphi_c$. The origin of this discrepancy, of course, stems from the very nature of the adopted structural model which explicitly assumes the existence of BI with altered heat conductivity around filler particles in the entire interval from $\varphi \geq 0$ to $\varphi \leq 1.0$. In other words, the hypothetical, ultimate state of a disperse phase at $\varphi \Rightarrow 1.0$ is the mosaic-like structure of InC composed of particles coated with BI at the highest packing density φ^*. This picture does seem physically reasonable as soon as there is no realistic mechanism for a disperse phase above $1 - \varphi_c$ to be converted into a monolithic, void-free body

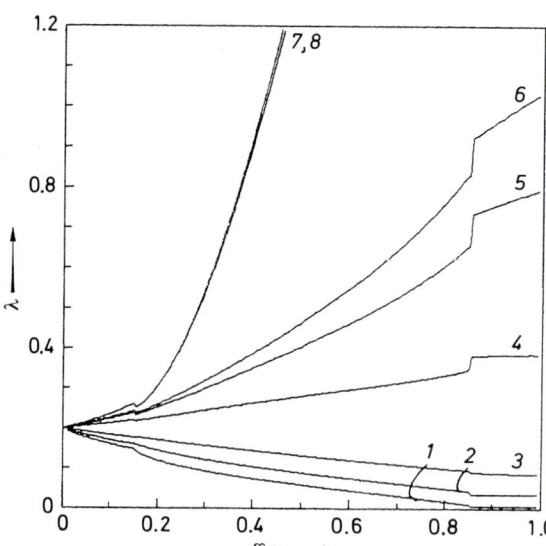

Fig. 11. Composition dependence of the heat conductivity of MHM for case ii) calculated assuming $\lambda_{BI}/\lambda_2 = 0.1$ and $\Delta l/r = 10^{-1}$ (curve 1), 10^{-2} (curve 2), 10^{-3} (curve 3); $\lambda_{BI}/\lambda_2 = 1$ and $\Delta l/r = 10^{-1}$ (curve 4), 10^{-2} (curve 5), 10^{-3} (curve 6); $\lambda_{BI}/\lambda_2 = 10$ and $\Delta l/r = 10^{-1}$–10^{-5} (curves 7 and 8)

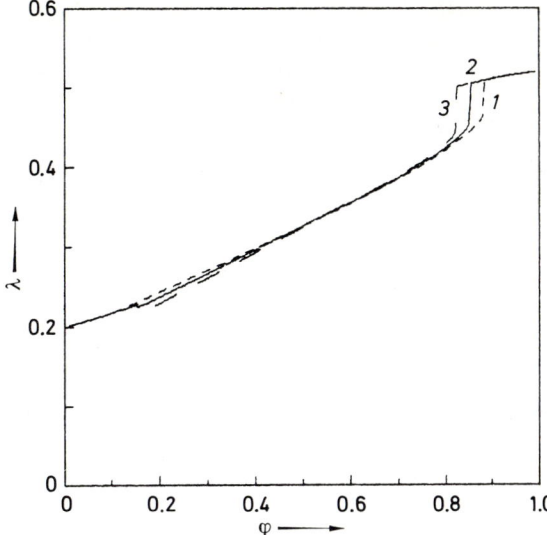

Fig. 12. Composition dependence of the heat conductivity of MHM for case iii) calculated assuming $\varphi_c = 0.12$ (*curve 1*), 0.15 (*curve 2*) and 0.18 (*curve 3*)

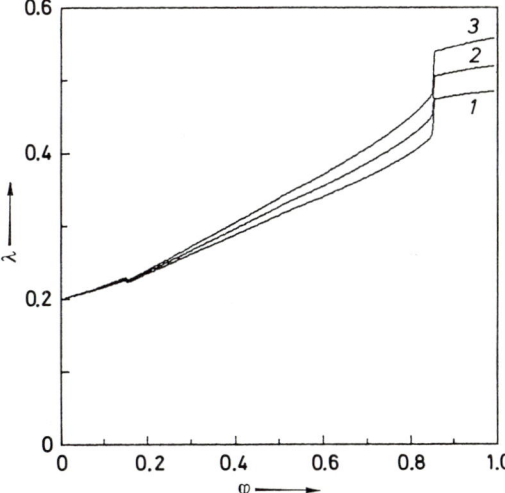

Fig. 13. Composition dependence of the heat conductivity of MHM for case iv) calculated assuming $\varphi^* = 0.4$ (*curve 1*), 0.6 (*curve 2*) and 0.8 (*curve 3*)

with intrinsic conductivity λ_1. Hence it is no surprise that the only case when the heat conductivity of MHM, λ, is predicted to increase from the lower theoretical limit ($\lambda_2 = 0.2$ W/m.K at $\varphi = 0$) to the upper theoretical limit ($\lambda_1 = 2.0$ W/m.K at $\varphi = 1.0$) corresponds to $\lambda_{BI}/\lambda_2 = \lambda_1/\lambda_2$ (curve 5 in Fig. 10). The other important observation is that the ultimate value of λ at $\varphi = 1.0$ becomes closer to λ_1 the lower the $\Delta l/r$ ratio (cf. curves 4, 5, 6 in Fig. 11). It is thus likely that the fitting parameters Δl and λ_{BI} are, in fact, interrelated, rather than mutually independent.

It should be recognized, however, that in view of the obvious experimental limitations, the pattern of λ evolution in the composition gap between $\varphi = 1 - \varphi_c$ and $\varphi = 1$ seems to have little practical importance. Therefore, $1 - \varphi_c$ may be regarded as a natural upper composition limit for the quantitative applicability of SSA technique.

4.2 Experimental Test

4.2.1 Filled Polymers

Heat conductivity of composite materials is notorious for its extreme sensitivity to even the smallest structural defects like binder-free voids between disperse particles which may occur, say, in samples prepared by mechanical mixing and subsequent hot-pressing of granulated polymers and inorganic powders [63, 64]. Thus theoretical predictions should be tested only on those samples which were prepared in conditions ensuring the elimination of such hazards.

It turned out, however, that in the majority of documented cases (e.g., [62–76]) the discrepancy between the experimental data obtained in routine heat conductivity measurements and the predictions of "pragmatic" approaches [e.g., Eqs. (19), (20), etc.] increased the higher the filler content φ, even for apparently defect-free samples. As a first guess, this discrepancy might have originated from the neglect of a "particle crowding" effect at higher φs which could however, be explicitly accounted for semi-empirically by "physical" Eq. (105) [77],

$$\lambda/\lambda_2 = \frac{1 + AB\varphi}{1 - B\Psi\varphi}, \tag{105}$$

where $B = (\lambda_2/\lambda_1 - 1)/(\lambda_2/\lambda_1 + A)$, $\Psi = 1 + [(1 - \varphi^*)/\varphi^{*2}]\varphi$ and A is the appropriate geometrical factor. In fact, the quality of the data fit to Eq. (105) compared to Eq. (19) or Eq. (20) could be considerably improved by a proper selection of the "best" values of parameters A and φ^* (maximum packing fraction of the filler) [77–79].

More extensive tests of predictions of different rigorous theories by several sets of experimental heat conductivity data obtained over broad intervals of compositions, temperatures and/or pressures [54, 80] will be illustrated below.

4.2.1.1 Filled Cross-Linked Epoxies [81, 82]

The test specimens were prepared by adding a desired amount of the filler (powders of non-metals [81] and metals [82] with particles sizes between 20 and 150 microns) to the base epoxy resin/hardener formulation which was subsequently cured at 373 K (1 h) and post-cured at 453 K (2 h). The heat conductivity of bar-shaped and disc-shaped specimens was measured in the temperature intervals 2–20 K and 20–300 K respectively.

In the temperature interval below 10 K the acoustic mismatch of heat-conducting phonons at the polymer-filler interfaces manifested itself as $\lambda < \lambda_2$, whereas at higher temperatures the "normal" increase of $\lambda > \lambda_2$ with φ was observed. The composition dependence of λ at all temperatures above 10 K was claimed to be adequately represented by Eq. (23) [81, 82], although the agreement with the SSA theory was apparently better, especially at the highest filler loadings (cf. the broken and the solid lines, respectively, in Fig. 14).

4.2.1.2 Polyethylene (PE)/NaCl [83]

Cylindrical specimens with filler volume content from $\varphi = 0$ (pure PE) to $\varphi = 1.0$ (pure NaCl) were prepared by a thorough mechanical mixing of components (both were powders with an average particle size of 200–300 microns) and subsequent compacting of mixtures at room temperature. (To ensure good thermal contact between the components and also to minimize the hazards of residual voids, the lower limit of compacting pressure was far in excess of indentation hardness of both PE and NaCl.) The heat conductivity was measured (estimated experimental inaccuracy about 4%) by the transient hot-wire method [84] in the temperature interval 120–320 K and in the pressure range 0.2–1.75 GPa.

As can be seen from a typical isothermal-isobaric plot of λ vs φ (Fig. 15), the quality of fit between the experimental data and the theoretical curve constructed according to the SSA technique with the set of parameters listed in Table 4

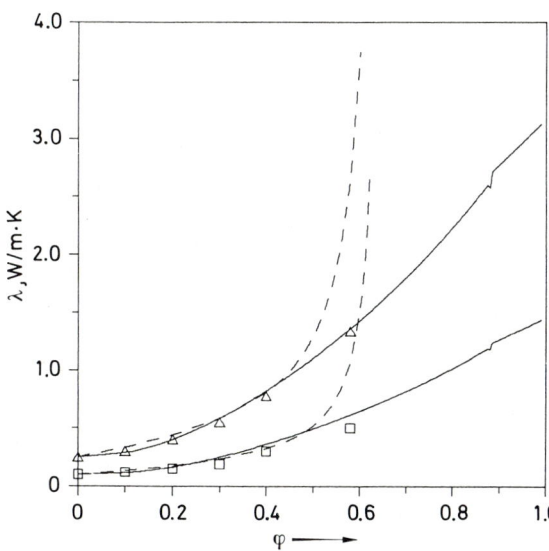

Fig. 14. Composition dependence of the heat conductivity of the copper-filled epoxy at 20 K (*squares*) and 300 K (*triangles*). *Broken curves* were calculated by Eq. (23), *solid curves* by SSA model assuming $\lambda_2 = 0.1$ W/m/K, $\lambda_1 = 5200$ W/m/K, $\varphi_c = 0.12$, $\varphi^* = 0.3$, $\Delta l/r = 0.017$, $t = 1.6$, $\lambda_{BI}/\lambda_2 = 3.5$ (20 K) and $\lambda_2 = 0.25$ W/m/K, $\lambda_1 = 395$ W/m/K, $\varphi_c = 0.12$, $\varphi^* = 0.55$, $\Delta l/r = 0.01$, $t = 1.5$, $\lambda_{BI}/\lambda_2 = 2.5$ (300 K)

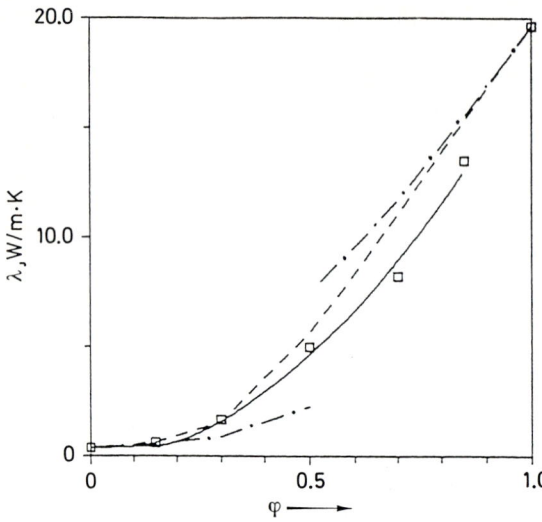

Fig. 15. Effective heat conductivity of PE/NaCl at 120 K and 0.2 GPa and predictions of SSA model (*solid line*), Eq. (17) (*broken-dotted line*) and Eq. (18) (*broken line*)

Table 4. Parameters of SSA model for filled PE (heat conductivity in [W/m/K])

Filler	λ_1	λ_2	λ_{BI}/λ_2	t	φ_c	φ^*	Δr
NaCl							
P = 0.2 GPa							
T = 120 K	0.371	19.61	39	1.6	0.15	0.49	0.01
T = 320 K	0.393	5.76	12.5	1.6	0.15	0.43	0.01
P = 1.73 GPa							
T = 120 K	0.582	26.54	37	1.6	0.15	0.49	0.01
T = 320 K	0.725	7.95	10	1.6	0.12	0.70	0.01
AgCl							
P = 0.11 GPa							
T = 120 K	0.353	2.26	5.7	1.6	0.12	0.54	0.01
T = 320 K	0.364	0.84	2.2	2.0	0.12	0.39	0.01
P = 1.75 GPa							
T = 120 K	0.582	3.12	5.6	1.6	0.12	0.50	0.10
T = 320 K	0.725	1.12	1.5	1.9	0.12	0.33	$< 10^{-3}$
Quartz (filler size 2r in microns)							
2r = 16	0.300	9.6	4.5	2.0	0.13	0.49	0.01
2r = 156	0.300	5.9	6.0	1.9	0.14	0.50	0.01
Al_2O_3	0.300	8.0	13	2.0	0.17	0.53	0.01
Graphite	0.300	209	10.5	1.6	0.12	0.50	0.01

(solid line) looks better compared to predictions of Eq. (94) and Eq. (105) (broken and dash-dotted lines, respectively). Essentially similar results were obtained at other pressures and/or temperatures (Fig. 16 may serve as a representative example).

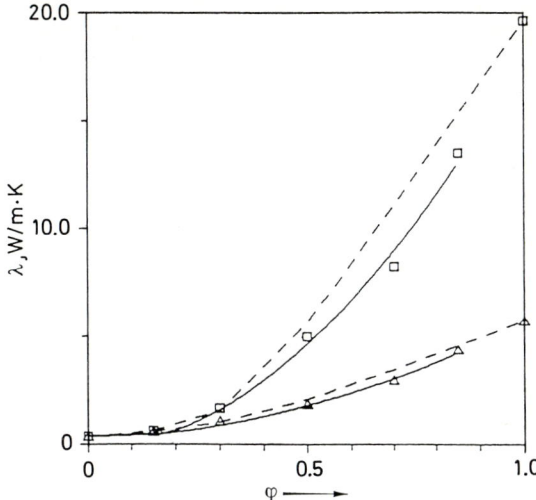

Fig. 16. Effective heat conductivity of PE/NaCl at 0.2 GPa and 120 K (*squares*) and 320 K (*triangles*) and predictions of SSA model (*solid lines*) and Eq. (18) (*broken lines*)

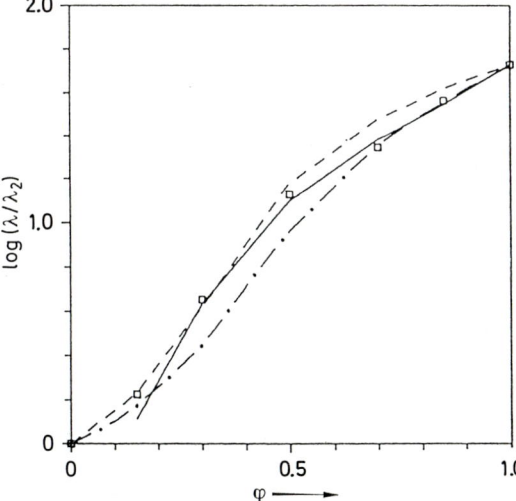

Fig. 17. Composition dependence of the ratio λ/λ_2 of PE/NaCl at $\lambda_1/\lambda_2 = 52$ (*squares*) and predictions of SSA (*solid line*), EMA (*broken line*) and RSRG (*dash-dotted line*)

In view of different dependencies of heat conductivities of both polymer and filler on temperature and pressure, a more stringent test of theoretical predictions would provide either a semi-log plot of the ratio, λ/λ_2 vs φ (Fig. 17) or rather a double-log plot of λ/λ_2 vs λ_1/λ_2 (Fig. 18). Again, the experimental data seem to be in a better agreement with the predictions of the SSA approach (solid lines) than with those of either EMA approach, Eq. (94), or RSRG approximation (broken and dash-dotted lines, respectively).

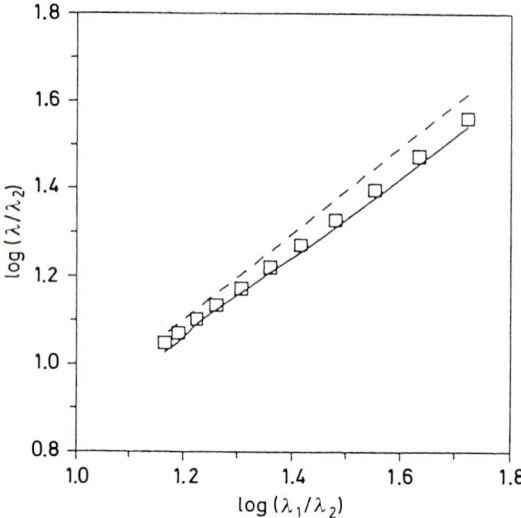

Fig. 18. Dependence of λ/λ_2 on λ_1/λ_2 for PE/NaCl at $\varphi = 0.85$ (*squares*), and predictions of the SSA (*solid line*) and EMA (*broken line*) approaches

4.2.1.3 Polyethylene (PE)/AgCl [85]

The instrumentation for heat conductivity measurements as well as the method of specimen preparation were essentially the same as in [83] except for a slightly broader temperature interval of measurements (100–300 K) and the lower initial pressure (0.1 GPa).

As was the case with the system PE/NaCl, application of the SSA technique with the set of parameters listed in Table 4 proved superior to other approaches.

4.2.1.4 Polyethylene (PE)/Quartz; PE/Al$_2$O$_3$; PE/Graphite [86, 87]

Disc-shaped specimens with filler volume content up to 0.8 were prepared either by melt-casting (MC) or by compression molding (CM) of binary mixtures of powdery components (PE with an average particle size of 10 microns, two quartz fractions with average particle sizes of 16 and 156 microns, two fractions of Al$_2$O$_3$ with particle sizes of 9 and 65 microns, and graphite with particles of unspecified size). Heat conductivity was measured at 323 K and normal pressure. The erratic behavior of λ for MC specimens above a (particle size-dependent) "critical" filler content indicated the onset of macrovoids formation, while such effects were not observed in the case of CM specimens for which the values of λ smoothly increased with φ.

Judged by the representative data in Figs. 19 and 20, the theoretical curves constructed according to the SSA approach with the fitting parameters shown in Table 4 (solid lines) adequately describe the experimental data for CM specimens up to the highest filler content, the quality of the fit being at least

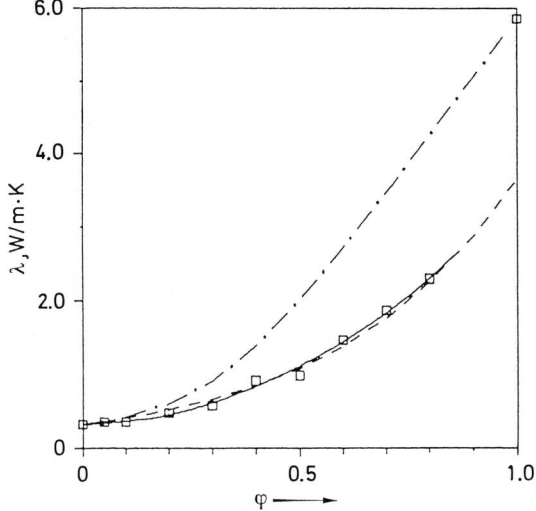

Fig. 19. Effective heat conductivity of PE/quartz (*squares*), and predictions of the SSA (*solid line*), EMA (*dash-dotted line*) approaches and Eq. (106) (*broken line*)

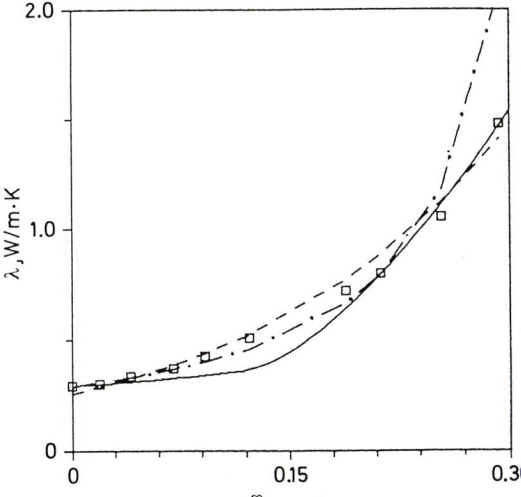

Fig. 20. Effective heat conductivity of PE/graphite (*squares*) and predictions of the SSA (*solid line*), EMA (*dash-dotted line*) and Eq. (106) (*broken line*)

comparable to that observed for the following semi-empirical relationship [86] (broken lines):

$$\log \lambda = \varphi C_1 \log \lambda_1 + (1 - \varphi) C_2 \log \lambda_2, \qquad (106)$$

where C_1 and C_2 are the fitting factors claimed to account for the probability of either aggregation of filler particles into conductive chains or morphological changes in the polymer phase in filled samples, respectively. (The broken lines in Figs. 18, 19 were constructed by Eq. (106) assuming $1.0 < C_1 < 1.5$ and $C_2 \cong 1.0$ [86].)

4.2.2 Polymer Blends

The experimental data are scarce; moreover, the composition dependence is not always readily amenable to a meaningful theoretical analysis as soon as the heat conductivities of the components are of the same order of magnitude (especially when both components are non-crystalline). As a typical example, the heat conductivity in the melt state (i.e., in the temperature interval 400–490 K) of melt-blended polystyrene (PS)/polycarbonate (PC) pair, estimated by substituting the experimental values of thermal diffusivity a, specific volume v and heat capacity C_p into the standard relationship, $\lambda = C_p a/v$, decreased approximately linearly with composition from $\lambda_2 = 0.32$ W/m.K for PC to $\lambda_1 = 0.25$ W/m.K for PS, as might have been expected for incompatible components [88, 89]. However, the negative deviations from linear additivity observed at low contents of either component were attributed to a limited mutual solubility of PC and PS.

Another example of an incompatible polymer pair is provided by melt blended specimens of oligomeric PE ($\lambda_2 = 0.34$ W/m.K) and PS ($\lambda_1 = 0.16$ W/m.K) containing a fixed small amount of a compatibilizing additive SEBS (PS/polybutadiene/PS triblock copolymer) [90]. The room-temperature heat conductivity vs composition plots (Fig. 21) could be empirically represented by Eq. (106) with two sets of parameters for PE contents below and above the apparent "phase inversion" point, respectively (broken line in Fig. 21). In contrast, only a single set of parameters ($\Delta l/r = 0.017$, $\varphi^* = 0.4$, $\lambda_{BI}/\lambda_2 = 2.5$) was required to fit the same data to the SSA theory (solid line in Fig. 21). The

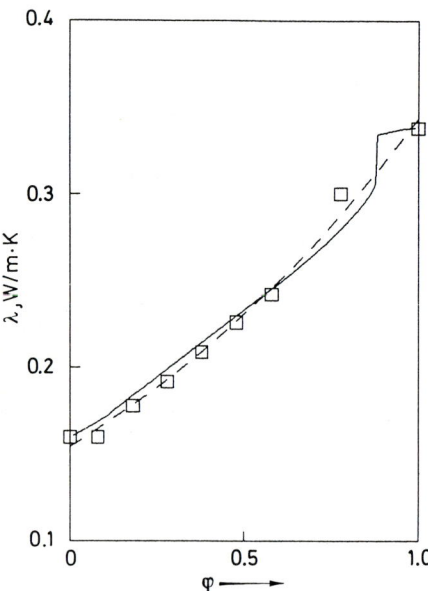

Fig. 21. Composition dependence of the heat conductivity of PE/PS blend (*squares*) and predictions of Eq. (106) (*broken line*) and of the SSA model (*solid line*)

same conclusion could be made after a similar treatment of the data on heat conductivity of PE/SEBS and PS/SEBS blends [91].

5 Conclusions

As follows from the detailed analysis of the current theories of heat conductivity in MHM, the "pragmatic" approaches neglecting the structural features of the latter are in approximate agreement with the experimental data at fairly small contents of the disperse component, whereas the "physical" approaches which account explicitly for the existence of BI and/or the occurrence of percolation phenomena are applicable over the entire composition range. Among the latter approaches, the SSA technique was shown to describe quantitatively the experimental data on composition dependence of the heat conductivity of various polymer materials over much broader composition interval (i.e., from φ_c to $1 - \varphi_c$) compared to EMA or RSRG methods. Moreover, the critical exponent for conductivity t which ensured the optimum fit of SSA predictions to the experimental data (Table 4) was also reasonably close to the corresponding theoretical estimates. Perhaps the quality of the data fit to theoretical predictions might have been improved even further, provided φ_c and φ^* were treated as fitting parameters, rather than fixed, universal constants. In fact there is experimental evidence suggesting that φ_c changes with polymer-filler interfacial energy and/or polymer melt viscosity [92, 93] whereas φ^* depends on particles shape [79]. Although, as follows from model calculations (Figs. 11 and 12), the exact value of either of these two parameters appears to be of secondary importance, the product $\varphi_c \varphi^*$ is of value as a theoretical measure of the minimum filler content (φ_{min}) for the onset of InC in real MHM. In this respect, $\varphi_{min} = \varphi_c \varphi^*$ of SSA approach is conceptually similar to the critical volume fraction of a conducting polymer, $\varphi_{min} = \varphi_c \varphi_c = \varphi_c^2$, invoked to account for the "double percolation" phenomenon in phase-separated, binary polymer blends [94].

It should be stated at this point that the basic assumption of structural "symmetricity" of each component near the corresponding percolation threshold (i.e., of a topological equivalence of component 1 at φ_c and of component 2 at $1 - \varphi_c$) implicit both in SSA technique and in conceptually similar RSRG approach, may well be an oversimplification of the real situation [63]. The other point of concern is the ratio λ_{BI}/λ_2, derived from curve fitting procedures for filled polymers, which may sometimes be of the order of λ_1/λ_2 itself. Taking into consideration that the heat conductivity of bulk, isotropic polymers is orders of magnitude smaller than the intrachain contribution, $\lambda_\parallel$, but comparable to the interchain contribution, $\lambda_\perp$ [95, 96], the observed inequality, $\lambda_{BI} \gg \lambda_2$, would imply that the structure of BI is radically different from unperturbed mother polymer phase. This hypothetical, higher-conductivity structure of BI might have involved chain orientation, somehow providing for the increased contribution from $\lambda_{||}$ [97, 98], and/or better interchain packing favoring heat conductivity in exactly the same way as does the densification under high pressure [99–101]. Lacking a quantitative, physically realistic model of BI, it is preferable at the present stage to regard both Δl and λ_{BI} as fitting variables of the SSA approach, rather than real "material" parameters of each particular two-component polymer system [80].

6 References

1. In the present context, the term "structureless" simply implies absence of macrodefects in both components (e.g., cracks between impinged spherulites in semi-crystalline polymers, or polymer-free voids between filler particles in filled polymers), perfect adhesion on the interface between components being assumed
2. Lipatov YuS (ed) (1986) Physical Chemistry of Multi-Component Polymer Systems. Naukova Dumka, Kiev, vols 1 and 2 (in Russian)
3. Deryaguin BV, Churayev NV, Myuller VM (1987) Surface Forces. Nauka, Moscow, 399 p (in Russian)
4. Lipatov YuS (1977) Physical Chemistry of Filled Polymers. Khimia, Moscow, 303 p (in Russian)
5. Hence, Δr is, in fact, not so much an intrinsic property of each combination of two components, as rather an "effective parameter", the numerical value of which will also heavily depend on the sensitivity and accuracy of the experimental technique used to measure the property of interest.
6. Kirkpatrick S (1973) Percolation and conduction. Rev Mod Phys 45(4): 574–582
7. Shklovsky BI, Efros AL (1979) Electronic Properties of Semi-Conducting Alloys. Nauka, Moscow, 235 p (in Russian)
8. Dulnev GN, Zarichnyak YuP (1974) Thermal Conductivity of Mixtures and Composites. Energhia, Leningrad (in Russian)
9. Misnar A (1968) Thermal Conductivity of Solids, Liquids, Gases and Their Mixtures. Mir, Moscow (in Russian)
10. Shermergor TD (1977) Theory of Elasticity of Microheterogeneous Media. Nauka, Moscow (in Russian)
11. Khoroshun LP, Maslov BP (1980) Computer Techniques for Calculation of the Physical Mechanical Properties of Composites. Naukova Dumka, Kiev (in Russian)
12. Hashin Z (1983) Analysis of composite materials. J Appl Mech 50: 481–505
13. Maxwell JC (1873) Treatise on Electricity and Magnetism. Oxford University Press, Oxford
14. Lorenz L (1880) Ueber die Refraktion Konstante. Wied Ann 11: 70–75
15. Lorentz HA (1880) Ueber die Beziehung zwischen der Fortpflanzungsgeschwindigkeit des Lichtes und der Koerperdichte. Wied Ann 9: 641–648
16. Malyshev VA, Malyshev PV (1987) Estimation of the effective dielectric permittivity of a periodic array of solid bodies. Dokl Acad Nauk UkrSSR, ser A,N 12: 48–53
17. Rayleigh LM (1882) On the influence of obstacles arranged in rectangular order upon the properties of a medium. Phil Mag 34: 19–22
18. Meredith RE, Tobias CW (1960) Resistance to potential flow through cubical array of spheres. J Appl Phys 31: 1270–1273
19. Bruggeman DAG (1935) Berechnung verschiedener physikalischer Konstanten von heterogenen Substanzen. Ann Phys 24: 636–650
20. Odelevsky VI (1951) Calculation of the effective conductivity of heterogeneous systems. Zhurn Tekh Fiz 21: 667–685
21. Kondorsky EK (1951) On the theory of coercetive force and magnetic susceptibility of ferromagnetic powders. Dokl Acad Nauk USSR 80: 197–200
22. Weinberg AK (1966) Magnetic susceptibility, electric conductivity, dielectric permittivity and heat conductivity of media with spherical and ellipsoidal inclusions. Dokl Acad Nauk USSR 169: 543–547

23. Fricke HA (1924) A mathematical treatment of the electric conductivity of disperse systems. Phys Rev 24: 12–15
24. Webman I, Jortner Z, Cohen MN (1976) Electronic transport in alkalitungsten bronzes. Phys Rev 13: 213–224
25. Webman I, Jortner Z, Cohen MN (1975) Numerical simulation of electrical conductivity in microscopically inhomogeneous materials. Phys Rev ser B 11: 2885–2892
26. Buyevich YuA, Korneev YuA, Shelikova IN (1975) On the heat and mass transfer in a disperse flux. Inzh Fiz Zhurn 30: 979–982
27. Buyevich YuA, Korneev YuA (1976) Effective heat conductivity of a disperse medium at low Pekle numbers. Inzh Fiz Zhurn 31: 607–612
28. Esksler BS (1979) On the effective heat conductivity and viscosity of disperse media. Inzh Fiz Zhurn 37: 110–117
29. Buyevich YuA (1973) On the effective heat conductivity of grainy materials. Zhurn Prikl Matem Tekhn Fiz N 4: 57–66
30. Davis RH (1986) The effective thermal conductivity of a composite material with spherical inclusions. Int J Thermophys 7: 609–620
31. Lichtenecker K, Rother K (1931) Die Herleitung des logarithmischen Mischungsgesetzes des allgemeinen Prinzipien der stationaren Stroemung. Phys Z 32: 3255–3267
32. Herring C (1960) Inhomogeneities in electrical and galvanometric measurements. J Appl Phys 31: 1939–1953
33. Kudinov VA, Moizhes BYa (1972) Effective conductivity of non-homogeneous, isotropic media. Zhurn Tekhn Fiz 42: 591–599
34. Hashin Z, Shtrikman SA (1962) A variational approach to the theory of the effective magnetic permeability of multi-phase materials. J Appl Phys 33: 3125–3131
35. Yermakov GA, Fokin AG, Shermergor TD (1974) Calculation of the bounds for effective dielectric permittivities of non-homogeneous dielectrics. Zhurn Tekhn Fiz 44: 249–254
36. Rosen BW, Hashin Z (1970) Effective thermal expansion coefficient and specific heat of composite materials. Int J Eng Sci 8: 157–161
37. Dykhne AM (1967) On the calculation of kinetic coefficients of media with random inhomogeneities. Zhurn Tekhn Fiz 52: 264–267
38. Kazantsev VP (1979) Variational estimates of the effective conductivity of media with macroscopic inclusions. Izv VUZ'ov, Fiz N 5: 53–59
39. Stepanov SV (1970) On the heat conductivity of two-phase systems. Inzh Fiz Zhurn 18: 247–252
40. Beran M, Molyneux L (1965) Boundaries of effective properties of binary heterogeneous disordered systems. Quart Appl Math 24: 107–114
41. Phan-Thien N, Milton GW (1982) New bounds on the effective thermal conductivity of N-phase materials. Proc Roy Soc ser A 380: 333–348
42. Levin VM (1968) On the estimates of effective conductivity coefficients of multi-phase materials. Prikl Mekh Tekhn Fiz N 2: 52–55
43. Novikov VV (1986) Two-bounds estimates of heat and electrical conductivity of microheterogeneous materials. Inzh Fiz Zhurn 50: 866–867
44. Hardi GG, Littlewood DE, Polya G (1948) Inequalities. Gosizdatinlit, Moscow (in Russian)
45. Kolmogorov AN, Fomin SV (1981) Elements of the Theory of Functions and of Functional Analysis. Nauka, Moscow (in Russian)
46. Uitrin AI, Belousov VYa (1985) Probabilistic-statistical methods of calculation and optimization of structural parameters of microheterogeneous composite materials. Poroshkov Metallurghia N 3: 69–73
47. Jackson JL (1968) Transport coefficients of composite materials. J Appl Phys 39: 1733–1736
48. Corriell SR, Jackson JL (1968) Bounds on transport coefficients of two-phase materials. J Appl Phys 39: 2329–2334
49. Hill R (1965) A self-consistent mechanics of composite materials. J Mech Phys Solids 13: 213–225
50. Frey S (1932) Ueber die elektrische Leitfaehigkeit binarer Aggregat Z Elektrochem 38: 260–274
51. Dulnev GN, Novikov VV (1977) Effective conductivity of systems with interpenetrating structures. Inzh Fiz Zhurn 33: 271–274
52. Novikov VV (1983) The effective thermal expansion coefficient of non-homogeneous materials. Inzh Fiz Zhurn 44: 969–977
53. Novikov VV (1985) On the estimates of effective elasticity moduli of non-homogeneous materials. Prikl Mekh Tekhn Fiz N 5: 146–153

54. Privalko VP, Novikov VV, Yanovsky YuG (1991) Principles of Thermophysics and Rheophysics of Polymer Materials. Naukova Dumka, Kiev (in Russian)
55. Hammersley LM (1983) Origins of percolation theory: in "Percolation Structures and Processes", ed. by Deutscher G, Zallen R, Adler J. Adam Hilger, Bristol, and the Israel Physical Society, Jerusalem, ch 2
56. Zallen R (1983) The Physics of Amorphous Solids. Wiley, New York-Chichester-Brisbane-Toronto-Singapore, ch 4
57. Stauffer D (1985) Introduction to Percolation Theory. Taylor & Francis, London and Philadelphia
58. Dulnev GN, Novikov VV (1979) Conductivity of non-homogeneous systems. Inz Fiz Zhurn 36: 900–909
59. Dulnev GN, Novikov VV (1983) Percolation theory and conductivity of non-homogeneous media. The basic model of a non-homogeneous medium. Inzh Fiz Zhurn 45: 136–141
60. Skal AS, Shklovsky VI (1974) Topology of an infinite cluster in the percolation theory and the theory of jump-like conductivity. Fiz Tekhn Poluprovodnikov 8: 1585–1592
61. Vinogradov AP, Sarychev AK (1983) The structure of conducting channels and the metal-dielectric transition in composites. Zhurn Eksper Teor Fiz 85: 1144–1151
62. Torquato S (1987) Thermal conductivity of disordered heterogeneous media from the microstructure. Rev Chem Eng 4: 151–204
63. Shah N, Ottino JM (1986) Effective transport properties of disordered, multi-phase composites. Application of real-space renormalization group theory. Chem Eng Sci 41: 283–296
64. Dulnev GN, Novikov VV (1979) On the conductivity of filled heterogeneous systems. Inzh Fiz Zhurn 37: 657–661
65. Novikov VV (1988) Influence of interphase layer and of structural parameters on the conductivity coefficient of polymer composites. Composites Polym Mater N 38: 17–23
66. Guyon E (1982) Percolation et matiere en grains. Compt Rend 294: 115–130
67. Ziman JM (1979) Models of Disorder. Cambridge University Press, London-New York-Melbourne
68. Novikov VV, Klimenko VS (1988) Heat conductivity of pseudoalloys. Tekhn Vysok Temper N 2: 10–13
69. Novikov VV, Dmitriev MV, Shapovalov IP, Tartakovskaya LN (1987) Estimation of the effective dielectric properties of glass-ceramics with account of structural parameters. Tekhn Sredstv Svyazi N 3: 264–269
70. Kusy R, Corneliussen R (1975) The thermal conductivity of nickel and copper dispersed in poly(vinyl chloride). Polym Eng Sci 15: 107–112
71. Kanari K (1977) Thermal conductivity of composite materials. Kobunshi 26: 557–561
72. Kline DJ (1961) Thermal conductivity studies of polymers. J Polym Sci 50: 441–450
73. Sundstrom DW, Lee Y-D (1972) Thermal conductivity of polymers filled with particulate solids. J Appl Polym Sci 16: 3159–3167
74. Hansen D, Tomkiewicz R (1975) Heat conduction in metal-filled polymers: the role of particle size, shape and orientation. Polym Eng Sci 15: 353–356
75. Privalko VP (1983) Thermophysical properties of filled polymers. Prom Teplotekhnika N 3: 66–76
76. Shut NI, Sichkar TG, Vozny PA (1985) Influence of the boundary layer structure on heat conduction and molecular mobility of filled epoxy systems. Composites Polym Mater N 24: 18–21
77. Nielsen LE (1974) The thermal and electrical conductivity of two-phase systems. Ind Eng Chem Fund 13: 17–20
78. Bigg DM (1979) Mechanical, thermal, and electrical properties of metal fiber-filled polymer composites. Polym Eng Sci 19: 1188–1192
79. Mamunya EP, Davidenko VV, Lebedev EV (1991) Properties of functionally filled polymer system in function of properties and content of disperse fillers. Composites Polym Mater N 50: 37–47
80. Rymarenko NL, Voitenko AI, Novikov VV, Privalko VP (1992) Composition-dependent properties of microheterogeneous polymeric materials: thermal conductivity of filled polyethylene. Ukr Polym J 1: 259–266
81. Garrett KW, Rosenberg HM (1974) The thermal conductivity of epoxy resin/powder composite materials. J Phys ser D (Appl Phys) 7: 1247–1258
82. de Araujo FFT, Rosenberg HM (1976) The thermal conductivity of epoxy resin/metal-powder composite materials from 1.7 to 300 K. J Phys ser D (Appl Phys) 9: 665–675

83. Hakansson B, Ross RG (1990) Effective thermal conductivity of binary dispersed composites over wide ranges of volume fraction, temperature and pressure. J Appl Phys 68: 3285–3292
84. Kutcherov V, Hakansson B, Ross RG, Backstrom G (1993) Experimental test of theories for the effective thermal conductivity of a dispersed composite. J Appl Phys 71
85. Hakansson B, Andersson P, Backstrom G (1988) Improved hot-wire procedure for thermophysical measurements under pressure. Rev Sci Instrum 59: 2269–2275
86. Agari Y, Ueda A, Tanaka M, Nagai S (1990) Thermal conductivity of a polymer filled with particles in the wide range from low to super-high volume content. J Appl Polym Sci 40: 929–941
87. Agari Y, Ueda A, Nagai S (1991) Thermal conductivities of composites in several types of dispersion systems. J Appl Polym Sci 42: 1665–1669
88. Yarema GE, Besklubenko YuD, Privalko VP (1982) Thermophysical properties of polystyrene/polycarbonate blends in the melt state under elevated pressures. Dokl Acad Nauk UkrSSR, ser B N 3: 54–57
89. Yarema GE (1993) Thesis, Institute of Macromolecular Chemistry, Academy of Sciences of Ukraine, Kiev, Ukraine
90. Agari Y, Ueda A, Nagai S (1992) Thermal conductivity of polyethylene/polystyrene blends containing SEBS block copolymer. J Appl Polym Sci 45: 1957–1966
91. Agari Y, Ueda A, Nagai S (1993) Thermal conductivities of blends of polyethylene/SEBS block copolymer and polystyrene/SEBS block copolymer. J Appl Polym Sci 47: 331–337
92. Sumita M, Asai S, Miyadera N et al. (1986) Electrical conductivity of carbon black filled ethylene-vinyl acetate copolymer as a function of vinyl acetate content. Colloid & Polym Sci 264: 212–217
93. Sumita M, Abe H, Kayaki H, Miyasaka K (1986) Effect of melt viscosity and surface tension of polymers on the percolation threshold of conductive-particle-filled polymeric composites. J Macromol Sci.-Phys ser B 25: 171–184
94. Levon K, Margolina A, Patashinsky AZ (1993) Multiple percolation in conducting polymer blends. Macromolecules 26: 4061–4063
95. Galeski A, Milczarek P, Kryszewski M (1977) Heat conduction in a two-dimensional spherulite. J Polym Sci: Polym Phys Ed 15: 1267–1281
96. Choy CL (1977) Thermal conductivity of polymers. Polymer 18: 984–1004
97. Choy CL, Luk WH, Chen FC (1978) Thermal conductivity of highly oriented polyethylene. Polymer 19: 155–162
98. Godovsky YuK (1982) Thermophysics of Polymers. Khimia, Moscow (in Russian)
99. Barker RE, Jr, Chen RYS, Frost RS (1977) Influence of pressure and chemical structure on the thermal conductivity of vitreous poly(alkyl methacrylates) I. J Polym Sci: Polym Phys Ed 15: 1199–1210
100. Rekhteta NA (1990) Thesis, Leningrad Institute of Fine Optics and Mechanics, Leningrad, Russia
101. Privalko VP, Rekhteta NA (1992) Effect of pressure on the thermal conductivity of polymers. J Therm Anal 38: 1083–1102

Editor: Prof. Godovsky
Received: May 1994

Electron Behavior and Magnetic Properties of Polymer Nanocomposites

D. Yu. Godovski
Russian Research Center "Kurchatov Institute", 123182, Moscow, Russia

In this review article, an attempt has been made to describe the relatively new class of composite systems, polymer nanocomposites. The study of nanocomposites is determined by a number of anomalous properties, exhibited by both the nanoparticles themselves and the systems of such objects immersed in a polymer matrix.

The anomalous character of nanoparticle properties is determined by their medium position between continuous bulk and single atoms. Such particles between 10 to 1000 Å sometimes exhibit a number of quantum size effects that determine anomalous optical and magnetic properties.

The cooperative effects of composites with interacting nanoparticles is another distinctive feature of such systems. These effects occur at the so-called percolation threshold, where the particles begin having contact with one another, whereby the interparticle contacts increase with the increase in their number. The electronic, optical and magnetic properties of composites, which change with the changes in cluster structure, are also discussed in this review.

1	Introduction: Properties of Nanoscale Systems and Classification of Nanocomposites	81
2	Synthesis of Polymer Nanocomposites	82
3	Methods of Characterisation of Polymer Nanocomposites	85
4	Properties of Nanoparticles in Polymer Nanocomposites (Local Properties of Nanoparticles)	89
	4.1 Electron Behavior in Nanoparticles	89
	4.1.1 Discretization of Electron Energy Spectrum	89
	4.1.2 Electroneutrality of Nanoparticles	90
	4.1.3 Charge Confinement Effects in Nanoparticles	91
	4.1.4 Further Localization of Electrons on Trap States	94
	4.2 Optical Properties of Nanoparticles	96
	4.2.1 Optical Absorption	96
	4.2.2 Luminescence Spectra	97
	4.2.3 Giant Dipole Resonance in Nanoparticles	98
	4.3 Catalytic Properties of Nanoparticles	99
	4.4 Debye Length Limit in Nanoparticles	101
5	Properties of Polymer Nanocomposites (Cooperative Properties of Systems of Nanoparticles)	101

- 5.1 Conductivity in Polymer Nanocomposites 101
  - 5.1.1 Interparticle Charge Transfer 101
  - 5.1.2 Peculiarities in Percolation Behavior of Electroconductivity 103
  - 5.1.3 The Influence of Polymer Matrix on Charge Transfer   108
- 5.2 Optical Properties and Photoconductivity Behavior of Polymer Nanocomposites 110
  - 5.2.1 Dielectric Confinement Effect 110
  - 5.2.2 Non-scattering of Light 111
  - 5.2.3 Photoconductivity 113
  - 5.2.4 Nonlinear Effects 117
- 5.3 Magnetic Behavior of Composite Systems with Nanoparticles   118

6 **Conclusions** 120

7 **References** 120

1 Introduction: Properties of Nanoscale Systems and Classification of Nanocomposites

The increasing interest in nano-size systems (i.e. the systems of size in the range 0.5 ~ 100 nm (5 ~ 1000 Å) such as nanocomposites [1, 2], superlattices of quantum dots [3], small metal clusters [4], isle-like films [5], cermets [6] and Q-colloids [7, 8] taken by physicists and chemists in recent years seems to be determined by a number of distinctive properties of both nanoparticles themselves, as the transition form between single molecule and continuous bulk media, and the anomalous cooperative properties of systems consisting of such objects. A number of properties, such as size quantization [7], monodomainity [9], and superpara- and superferromagnetism [10], are unique and thus stimulate the interest of theorists and experimentalists to these systems.

Let us clarify the terminological confusion: quantum dots, Q-particles, clusters, highly disperse particles, nanocrystals – such a broad spectrum exists covering exactly the same objects. Using the classification in [11] we will call particles of sizes less than 10 Å clusters, and the objects with size greater than 10 Å, nanoparticles. This difference concerns not only the terminology; there exists real differences in the properties of objects with sizes below and above 10 Å. These arise from the ratio of crystalline lattice deformation. For particles of 10 Å or smaller size the lattice is usually deformed if it exists at all. For larger particles, numerous investigations have proved the lattice parameters to be only slightly different from those of the bulk state [7, 12, 13].

Polymer nanocomposites (i.e. the composites consisting of polymer matrix filled with nanosize particles) are especially interesting due to their high long-time stability (polymer matrix prevents both oxidation and coalescence), distinguishing them from thin solid films and free high-dispersion particles. They also have relatively simple synthesis methods.

Another point is that one can obtain not only "diluted" systems with low filler content, and can investigate the properties of noninteracting particles in polymer matrix but also highly filled composites with interacting nanoparticles. These can be synthesized and the cooperative behavior of interacting particles studied. The ability to obtain composites with threshold properties ranging from noninteracting to cooperative is another distinctive feature of such systems which will be discussed in Sect. 3. The peculiarity of polymer nanocomposites in comparison with other nano-size objects consists of the influence of polymer matrix on properties of the composite and of the interaction which may take place between the matrix and nanoparticle.

We would like then to distinguish the properties of nanoparticles themselves from cooperative properties exhibited by nanocomposites as ensembles of such small particles. The properties of nanoparticles themselves will be described in the Sect. 4 and the cooperative effects in Sect. 5 of this review. We also limit our observation to only the 0–3 nanocomposites [14], i.e. the systems in which filler is 0-dimensional (particle) and matrix is 3-dimensional and continuous. We will not cover here either the 1–3 composites (rod like 1-dimensional filler in 3-dimensional matrix) or the 2–3 composites or intercalates. The 0-3 composites, according to the classification referred to above [14], are divided into matrix mixtures in which the filler particles are ordered in superlattice with repeating interparticle distances and the so-called stochastic mixtures in which the particles are randomly dispersed in polymer media. There was no information in periodicals on polymer nanocomposites with strong matrix ordering. Since some properties of such systems could be very promising, being close to those of the superlattices, the existing synthesis methods do not provide the essential ordering of filler particles.

Thus the composites treated in this review are of stochastic type. It is necessary to mention, nevertheless, that the distribution of nanoparticles in polymer matrix is not completely random: the synthesis methods induce ordering in particle position and aggregation often takes place; thus it is often not a true stochastic distribution which is observed.

Hence the objects of this review are polymer composites filled with nanoparticles ($1 - 10^2$ nm). Due to the high reaction ability of such nanoparticles, the filler must be inert enough to ensure the stability of composites and this criterion is best met by sulfides – CdS [7], CuS[2], ZnS [15], PbS [16]; oxides – Fe_2O_3 [17], halogenides – PbI_2 [18], CuCl [19] and noble metals – Ag [20], Au, Pt [21].

2 Synthesis of Polymer Nanocomposites

In recent years, various methods of synthesis of nano-size particle systems have been established. They have used aqueous [22] and non-aqueous [23] solutions, reversed micelles [24], vesicles [24], BLMs [24], surface monolayers [25], clays [26], zeolites [27], Langmuir-Blodgett films [28], peptides [29] and yeast cells [30]. The main idea of the cited variety of synthesis methods is to control the particle size either by spatial conditions (e.g. the size of pores [29], entities in media [24, 27]) or kinetics of synthesis reaction [22]. Another parameter, which in many cases decreases the particle size, is temperature [2, 16].

In our review we will consider only methods of synthesis of polymer composites. Three main methods were established recently, by which such variety of polymer composites has been obtained. The in-situ method or the formation of particles in polymer matrix immersed in aqueous or non aqueous solution involves soluble reagents transported into polymer, fixed by matrix containing specific groups and then reacted with the formation of non-soluble particles; the by-products of reaction are extracted by washing. An example of such synthesis is the formation of CuS particles with the size of ~ 20 nm (SAXS) in polyvinyl alcohol matrix mixed with polyacrylic acid [2]. The initial reagents are $CuSO_4$ and Na_2S aqueous solutions in stoichiometric quantities. Acidic groups of polyacrylic acid serve as the complexation centers for Cu^{2+} ions (taking into account the basic character of solution – pH $\sim$ 12) and sulfide S^{2-} ions react with such centers with the formation of CuS microcrystals in accordance with the reaction:

$$CuSO_4 + Na_2S \rightarrow CuS + Na_2SO_4. \tag{I}$$

Na_2SO_4 is then extracted by water and the composite films are dried in absolute alcohol. Polyacrylic acid can also be extracted by multiple washing in water. Decrease in temperature leads to the decrease of particle sizes down to 10 nm. A number of non-soluble sulfides such as CuS, ZnS, CdS were synthesized in polyvinyl alcohol matrix by means of this method, which can also be applied to oxide formation [2]. The particle size in that case is determined by the deformation of polyvinyl alcohol matrix and, in first approximation, does not depend on filler concentration.

Another example of the in-situ method is presented in [31]. As a polymer matrix, ethylene-15% methacrylic acid copolymer (E-MAA) is used, providing for good mechanical and optical properties and kinetic stability of composite. The synthesis consists in milling E-MAA (neutralized ionomer particles of $\sim$ 5 nm size) with lead (Pb) acetate or acetylacetonate at $\sim$ 160 °C. Metal cations form with E-MAA the polar clusters, carboxylate groups of the copolymer, which surround the clusters serve as an anions. Such E-MAA films, 30–300 μm thick, containing 0.0005–1 M Pb^{2+} are exposed to 1 atm of H_2S in a temperature range 25–125 °C for at least two hours.

Colors of the finally obtained composites vary from yellow, orange, red, brown or black with increasing concentration of Pb^{2+}. The clusters of PbS were found to have a crystalline nature (WAXS) and the size, estimated using the Sherrer equation (see below) was from 2.5 to 5–7 nm. It was found that at 25 °C the particle size is determined by the initial loading of Pb^{2+} within each original cluster, but at 125 °C, migration and agglomeration within polymer matrix occurs. During the aggregation from 2.5 to 4.5 nm the color changes from orange to black. The activation energy of aggregation from 2.5 to 4.5 nm was estimated to be 76 kcal/mole and aggregation time at 25 °C by extrapolation was about 10^5 years. PbSe, ZnS, ZnSe and CdSe were obtained by the same method. The other variations of the in-situ method are the synthesis within the channels of perfluorocarbon sulfonic acid membranes [32] or synthesis within the crazes of polymer, stretched in organic solutions [33].

The second method of synthesis of nanosize composites is the polymerization of a colloid solution containing colloid particles of metals [21], sulfides [7], or hydroxides [34]. A number of composites were obtained by that method, such as CdS [7], CdSe [35] and ZnS [36]. The particle size can be controlled by reaction temperature and the properties of a colloid solution; thermal coagulation and Ostwald ripening can be controlled via double-layer repulsion of individual crystallites [13] by carrying out synthesis at low temperatures in non-aqueous solvents and by adsorption of foreign stabilizing molecules. For CdS and PbI_2

there is clear evidence for "magic numbers" in the initially prepared clusters (nanoparticles). The kinetics of nucleation has been studied [37, 38]. It is necessary to mention that colloid solutions containing small particles are also the objects of investigation of nanoparticles themselves. They are relatively stable and form clear organic glasses on cooling. Thus absorption, Raman and luminescence measurements can be made.

Another example of this method is the synthesis of dual Au/Pt composites – the aqueous solutions of polyvinyl alcohol and poly-(N-vinyl-2-pyrollidone) have been used [39] with polymerization in the solution after colloid particle formation. The aggregation of nanoparticles in this case does not usually take place; however the polymerization conditions affect the homogeneity of the composite obtained. They show some catalytic properties, characteristic of bimetallic catalysts. There were some reports on semiconductor composites with metal coatings such as CdS/Rh 40 Å composites [44] and CdSe/Ag 50 Å composites [45]. The former system exhibits enhanced photochemical properties, the latter was used for investigation of Raman scattering of nanocomposites.

The combination of the first and second methods is a reaction parallel with gel formation from solution [39]. Thus, by successive treatment of polyvinyl alcohol solution containing glutaric aldehyde as gel-forming component with the solutions of $CuSO_4$ and Na_2S, the composites can be synthesized since the reaction $CuSO_4 + Na_2S \rightarrow CuS + Na_2SO_4$ takes place simultaneously with gel formation giving composites with CuS particles of ~ 40 nm size [39].

Another method of synthesis mentioned in [39] is the direct mechanical mixing of a polymer solution (aqueous or non-aqueous) with initially synthesized highly dispersive particles of nanosize with vaporization of water from the solution. In this case the inhomogeneity of filler distribution is high and coagulation often takes place [39]. The nanoparticles are usually synthesized by this "arrested precipitation". The phosphides and arsenides of Zn and Cd [12], CuCl [12], semiconductor PbI_2 and Fe_3O_2 were reported to form nanoparticles in this way.

We will mention one promising method which can also be referred to as a preparation of polymer nanocomposites. This method consists of capping the surface of nanoparticle with organic or inorganic groups such that the system is stable against agglomeration [40]. Polyphosphates [40] and thiols [41] are the most commonly used capping agents. In the work of Dance et al. [42] a molecular fragment of sphalerite CdS containing a $Cd_{10}S_4$ core was capped by 16 thiophenolate groups. This synthesis was carried out in solution but the clusters could be collected as stable solids and redissolved back into solution.

Wang et al. [43] reported on the reaction between cadmium and chalkogenide ions as an example of inorganic polymerization with an initial step forming a small nucleus which grows larger in a propagation step before being stopped in a termination step. Using a sulfide as the propagation reagent and a thiophenolate as the termination reagent, and adjusting the concentrations, the thiophenolate-capped CdS clusters of varying sizes up to 40 Å were obtained in large quantities (up to tens of grams).

Let us consider the synthetic systems, which can be referred to as a new step in nanocomposites preparation – the systems, in which any particle consists of a number of layers. The double-layer particle systems CdS-HgS in colloid solutions [46] were studied and their optical behavior was investigated. It has been found that varying conditions of colloid solution (primarily the concentration of components) lead to the particles exhibiting two main structures: the CdS nuclei wrapped by HgS layer and HgS nuclei wrapped by CdS layer.

Further fascinating developments of such polylayer nanoparticles are proposed in the same paper, namely the three layered small particles formed by two semiconductor materials with various band gap widths: the nucleus of one material, the thin spheric layer of another semiconductor and the coating third layer of the nucleus material (for example, CdS-HgS-CdS). The small particles of such type are assumed to possess the properties of sphere-like quantum wells, with all the properties of 2-dimensional quantum objects, known as quantum wells in doped semiconductors. These examples illustrate recent developments in preparation methods for nanocomposites from simple to complicated systems, which can be compared with the products of nanotechnology in semiconductor microelectronics.

3 Methods of Polymer Nanocomposites Characterization

First and paramount method of nanocomposites characterization is, undoubtedly, transmission electron microscopy (TEM), which allows one to observe the particle shape, to determine particle sizes, to control the homogeneity of the composite and to obtain the histogram of nanoparticle sizes (Figs. 1, 2). TEM (especially axial bright field microscopy) is the most informative method of characterization with the resolution up to units of angstroms allowing visualization of the lattice plane and measurement of interplanar distances of nanoparticles. However, its application to polymer composites is hindered due to charge instability of some polymers (such as PVA [39], or PVDF [35]) and by the difficulty of sample preparation.

Another informative method of nanocomposites characterization is the wide angle X-ray structural analysis (WAXS) and especially small-angle X-ray analysis (SAXS). X-ray analysis allows one to determine the lattice parameters of nanoparticles and, as it was found [2, 12] for the majority of sulfides and oxides, the small particles exhibit the same lattice parameters and crystallographic structure as bulk substances up to the size of ~ 20 Å. At this size, the deformation of the lattice takes place which accompanies the transition to cluster structure with minimized full energy, including free energy of the surface [11].

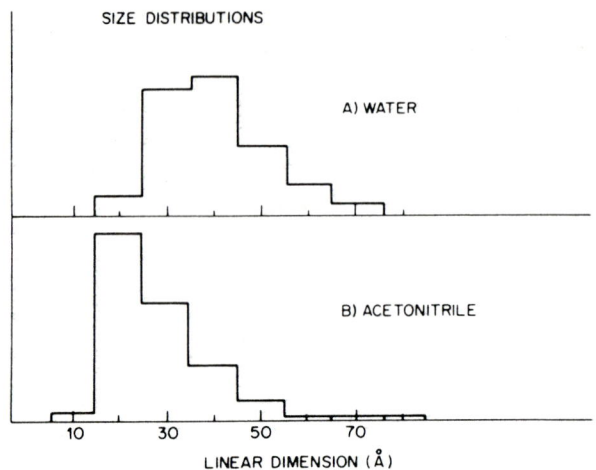

Fig. 1. Size histogram for fresh aqueous CdS colloid and acetonitryle colloid [7]

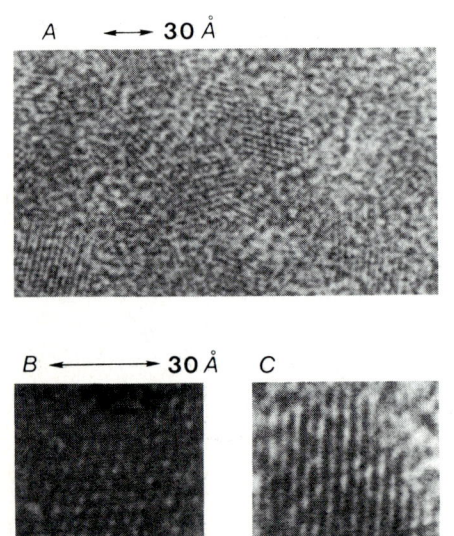

Fig. 2a–c. High resolution axial bright field transmission electron micrograph taken near Sherzer focus, for which the point-to-point resolution is = 2.5 Å. The sample consists of dispersed CdS crystallites on a thin carbon support film: **a** an area containing several crystallites and demonstrating the size variational **b, c** magnified images of individual crystallites [7]

The sizes of crystallites (which are not equal to nanoparticle size in the case of polycrystalline particles) can be determined using the Sherrer (Sherrer–Jones) expression for reflex width:

$$L = 0.9 \, (\lambda/b) \cos \theta \qquad (1)$$

where b is the reflex width on half of its height, θ – angle in radians in 2θ scale, and λ – wavelength of X-rays (usually 0.154 nm). The error of size determination is about 15% and is usually shifted to larger sizes.

Small-angle X-ray measurements (SAXS), another powerful method of nanocomposites characterization, is based on the analysis of diffraction in the range of 5–10 to 400–500 angle seconds, which corresponds to the inverse lattice modulus values $S = 0.0007$–0.426 ($S = 2\pi\Phi/\lambda$ for $\lambda = 0.154$ nm). To get rid of the effect of polymer matrix inhomogeneities it is necessary to subtract the scattering of a buffer, i.e. the matrix without particles [47]. The interference effects in spectrum often point to the association of small particles. The intensity on downfall of low-angle curves usually fits the Porod's law: $J \sim s^3$, where s is the angle in seconds. This fact reflects the homogeneity of nanocrystals, a sharp interface being assumed.

The mean inertia radii, mean volume and mean surfaces are calculated in accordance with the following expressions [47]:

$$\text{Ln } I(s) \cong \text{Ln } I(0) - s^2 R_g^2/3 \quad (sR_g < 1) \tag{2}$$

$$V_1 = 4\pi \int_0^D R^2 \gamma_o (R) dR \tag{3}$$

$$V_2 = 2\pi I(0)/Q_s^2 \tag{4}$$

$$Q = \int_0^{ud} S^2 I(S) dS + C_4/S_{od} \tag{5}$$

where I(s) is the value of reflection after subtracting the collimation (instrumental) coefficient, I(0) – the value of intensity extrapolated to 0, $\gamma_0(r)$ – related correlation function, D – max size, determined for $\gamma_0 = 0$, s_{od} – maximal value of s, for which the assumption of particle homogeneity is valid, C_4 – coefficient in the asymptotic expression: $I(s) = C_o + C_4/S^4$ for $s \to 0$. The calculation of volume ratio of nanoparticles w is conducted using the expression:

$$\langle \Delta \eta^2 \rangle = (\Delta \eta)^2 w(1-w), \tag{6}$$

where $\langle \Delta \eta^2 \rangle$ is the mean square variation of electron density. Nevertheless, it is often hard to determine the particle shape and distribution of particle sizes, especially in the case of a broad size distribution [47].

Recently, it has been shown that EXAFS is another powerful method, complimentary to SAXS [48]. It is determined by the local character of EXAFS analysis, which tests the first few shells of chosen atoms and allows one to determine such parameters as coordination numbers and distances for at least the first shell. Another important data, derived from EXAFS spectra analysis, are the Debye–Waller factor, which allows one to deduce some vibration properties and to determine the local stresses of the lattice.

In [48] the capped CdSe 40 Å-size particles were studied by EXAFS. It was found that the interior is bulklike, except for static strain, but the surfaces of nanoparticles are extensively reconstructed. The temperature dependence of the

Debye–Waller factor was found to be the same as that for bulk, except for addition of static contribution. Static strain for Cd-Se derived was ~ 0.015 Å, on average.

Another method of size determination [12] is based on resonant Rayleigh reflexion if the polymer matrix allows such experiments, i.e. is transparent in the range of a visible light. In this case, the determined sizes are shifted due to the influence of polymer, which affects the reflexion on the interface between a small particle and a polymer. Thus the sizes are usually larger than those determined by other methods. The overestimation error is usually of the order of 10–30% of the size value.

The other method of particle size determination, along with particle characterization, is Auger-microscopy [11]. It has been found, that the position of Auger peaks is shifted to the lower energy area with decrease in particles sizes (see Fig. 3). This shift is due to higher energy surface electron orbits compared to those in the bulk. The broadening of the Auger-peaks also points to the broadening of the energy spectrum of molecular orbits on the surface of nanoparticles.

Ionization Losses Spectroscopy is the other efficient method of small particles characterization. The initial electron ionizes the surface atom on some of the internal levels. The energy of a secondary electron is maximal if the condition: max $E_k = E_p - E_i$ (where E_p – initial electron energy, E_i – ionization

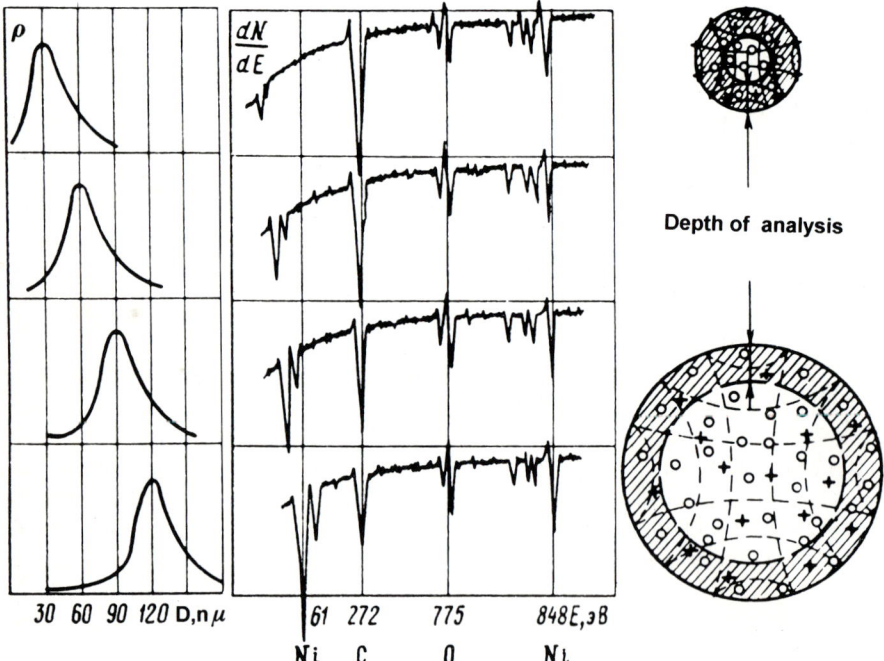

Fig. 3. Auger electron spectra dN/dE of nanosize nickel particles and corresponding functions of size distribution [11]

energy) is valid. These lines of the spectrum, corresponding to ionization losses are shifted with the shift of initial electron energy, contrary to nonshifting Auger peaks. Thus the ionization energy of electrons in small particles can be determined using this method.

4 Properties of Nanoparticles in Polymer Nanocomposites (Local Behavior of Nanoparticles)

The specific character of properties, demonstrated by nanocomposites is determined by the small size (units of nanometers) of filler particles, comparable with the wavelength of electron, which leads to the so called "quantum size effects" and the essential ratio of surface to volume in such systems, which increases the role of particle surface and interfaces between particle and polymer media (e.g., in a 50 Å CdS particle, about 15% of the atoms are on the surface). The latter fact is the reason for the higher chemical activity of nanoparticles and the increase in the role of such surface excitations as surface plasmons in small metal particles and specific surface phonon modes both as the increasing role of surface states, especially surface traps in semiconductor nanoparticles.

In this section, we will briefly describe the properties of small particles with the sizes of 5–100 nm. According to [11] the particles of size below 1 nm are referred to as "clusters", and the objects with sizes in the range from 1 nm to 1–10 μm are called "small" or "nanoparticles". We will primarily concentrate on electron behavior in small semiconductor and metal particles, but not in clusters.

The two main features distinguishing electron behavior in small particles are the size quantization with the discretization of electron energy spectrum, and the charge confinement, i.e. the inability of charge carriers to escape the particle, and the existence of boundary conditions for the electron density waves.

4.1 Electron Behavior in Nanoparticles

4.1.1 Discretization of Electron Energy Spectrum

Discretization in small particles is caused by their intermediate position between molecules, having a discrete spectrum of electron energy levels, and bulk solids, which have the quasi continuous spectrum of electron energy levels. Small particles comprising thousands or tens of thousands of atoms have distances between energy levels of the order of 10^{-4} eV – i.e. rather small but detectable.

The influence of discretization can be observed when the effect characteristic energy is of the same order with the distance between levels $\delta = 1/W(E_f) = 4/3$ E_f/N. Consider for example, the cubic particle with a size of 100 Å (N ~ 2.5 10^4 and $\delta = 1.5\ 10^{-4}$ eV or $k_B \times 1.3$ K); the effect of spectrum discretization can be observed at low temperature ($\delta > kT$), in weak magnetic fields ($\delta > \mu_B H$), or at low frequencies ($\delta > \hbar\omega$). These conditions are not common and thus the discretization effects are visible only in special cases. The anomalous behavior of paramagnetic susceptibility for small particles was predicted by Kubo for low temperatures, provided Zeeman energy is less than interlevel distances ($\mu_B H \ll \delta$). [see, for example [11]]. The difference in susceptibility for systems with odd and even numbers of electrons was predicted and observed experimentally (see Fig. 4). The anomalous linear dependence of low-temperature heat capacity on temperature was also predicted by Kubo in assuming that energy levels are stochastically distributed due to surface disorder of spectral properties. Figure 5 shows the Denton and Kubo dependencies along with experimental data. The discretization of energy levels must also affect the relaxation processes in small particles [11]. In continuous conducting media, Zeeman energy in relaxation processes is compensated by change of electron kinetic energy. In small particles this compensation hardly takes place. The Corringa mechanism for nuclei relaxation is also restricted in small particles. Thus the relaxation mechanisms, effective in continuous media, are noneffective in small particles and the relaxation mechanisms in them could be different and are still under investigation.

4.1.2 Electroneutrality of Nanoparticles

The other distinctive feature of nanoparticles is the electroneutrality condition. The work of charge extortion is equal to e^2/r. For the hydrogen atom it is the ionization energy (13.5 eV). If the radius is 50 Å or 500 Å, the ionization energy is 0.13 and 0.013 eV, respectively. Thus at ambient temperature ($kT \sim 0.025$ eV)

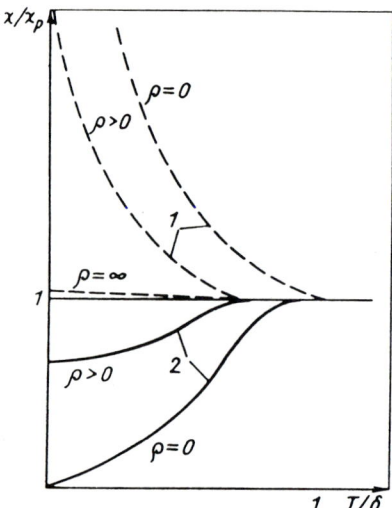

Fig. 4. Spin paramagnetism of metal clusters with odd (1) and even (2) numbers of electrons. As the spin–orbital interaction increases, the spin paramagnetism behavior approaches normal Pauly paramagnetism [11]

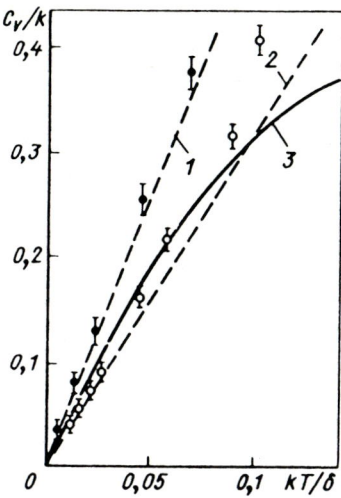

Fig. 5. Theoretical dependence of thermocapacity of metal nanoparticles on temperature: 1,2–data obtained by Kubo; 3–results of Denton, averaged for odd and even electron quantities; *circles and filled circles*–average values of 1000 spectra sets with even and odd electron numbers [11]

the 50 Å particle will remain neutral. The electron levels of small particles are not too broadened due to electron-phonon interaction as the distance of free motion is of the order of hundreds of angstroms, and the electron–electron interaction is restricted due to the Pauli principle and charge screening. Both of these reasons broaden the lines only by $\sim 10^{-6}$ eV at 4 K. Nevertheless, in the systems of contacting particles it has been reported, that the broadening could be caused by interparticle tunneling transitions [11].

4.1.3 Charge Confinement Effects in Nanoparticles

Another characteristic feature of electron behavior in nanoparticles is the charge confinement, i.e. the presence of the potential which holds the carriers in the particle and thus determines the maximal size of carrier motion ("particle in a box" model). It is the presence of a confinement but not the discretization caused by finite quantity of atoms in a particle effect (as described above), which is responsible for most of the so-called "quantum size effects." The effects referred to originate in the very fact, that the particle size is comparable to electron or hole De Broglie length and to form the spatially stable solution both finite size and boundary conditions are to be taken into account. In the majority of models the finite character of confinement border potential (and hence the possibility for the carrier to escape the particle or tunnel to another particle) has not been taken into account. In Sect. 5 of our review we will show that the variety of effects determining the cooperative behavior of highly-filled composites are caused by this possibility.

The simplest model in this case is the square-well confinement potential of infinite depth, well known from quantum physics courses. The specificity in this case consists of the fact that a small particle has the crystalline lattice of a bulk solid and, hence, the band structure and quasi-impulse spectrum of the lattice. Thus what is "trapped" into confinement is not a free particle but the excitation or a quasiparticle existing in the bulk lattice. The most fascinating fact is that the presence of a confinement drastically changes the properties of quasiparticles and hence the whole electron behavior of the system is changed.

Two main approaches can be used for description of electron states in nanoparticles. The first is the MO LCAO model due to transition properties of such objects between single molecule and bulk. The other method, originated from solid state physics, is the exciton theory, which is applied to small particles. It is important, because in the majority of cases the electron-hole pair is confined to the small particle and the exciton exists for a fairly long time.

Since the unit cell in small particles is the same as in the bulk materials, it points to the fact that the cluster MO evolve into Bloch MOs as the size increases. MO localized in space could be approximated as a Fourier transform wave packet of Bloch MOs:

$$\psi_i(r) = \sum_\mu \int_K f_{im}(k) \, \Phi_{km}(\bar{r}) dk. \tag{7}$$

Here μ is an index for summation over different bands. The band structure of a cubic crystalline CdS is depicted in Fig. 6.

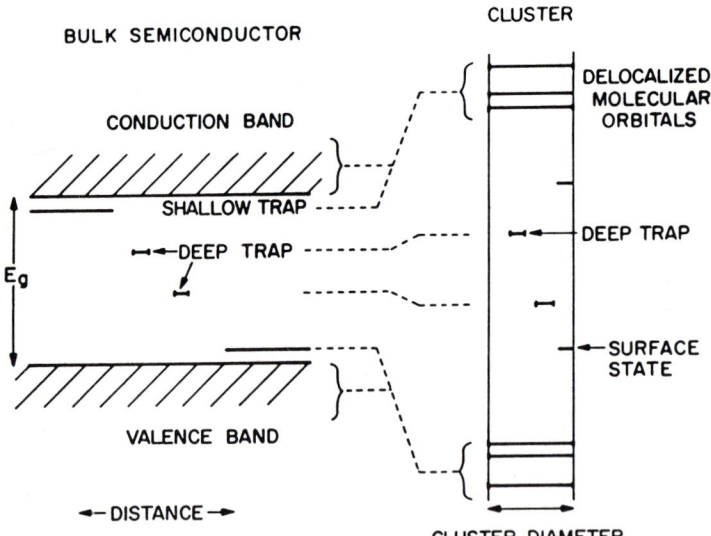

Fig. 6a. Schematic diagram relating cluster states to bulk crystal states adapted from [13]. **b** Schematic band structure diagram of cubic crystalline CdS. *The lines labeled L* (111) *and X* (100) *refer to* different directions within the unit cell [12]

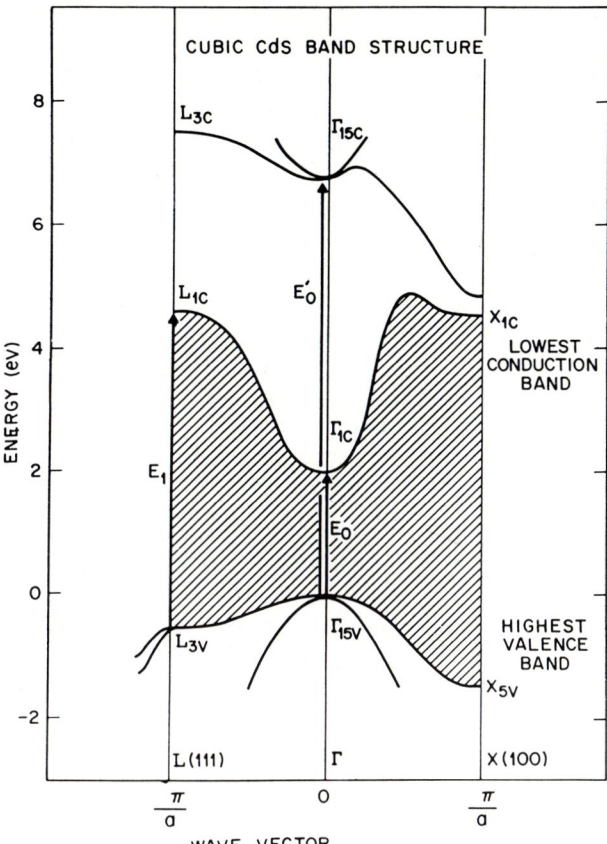

Fig. 6 (Continued)

The smaller the particle the larger the region of k space necessary to localize the electron inside the cluster. Size dependent expectation value of electron energy is:

$$E_i = I\langle\psi_i|H|\psi\rangle I \sim E_c + \frac{\pi^2}{2R^2} \sum_{i=x,y,z} \left\{\frac{d^2E}{d_{ki}^2}\right\}. \tag{8}$$

The expression includes only the lowest order nonzero term in Taylor's series (assuming the spherical shape of particles). The effective mass tensor is defined as:

$$M_{ij} = h^2 \left/ \left\{\frac{d^2E}{d_{ki}^2}\right\}\right. . \tag{9}$$

Near the k = 0 point this tensor is isotropic with average diagonal element m_e. Thus we obtain:

$$E_i = E_c + \frac{\pi^2 h^2}{2m_e R^2} \tag{10}$$

The second term corresponds to particle-in-a-box quantum energy. One can say that the presence of a confinement potential shifts the energy of electrons and holes in accordance with Eq. (10). An analogous approach in MO LCAO terms gives the similar results (see [13]). We obtain the electron energy at LUMO (the single electron energy near the bottom of conduction band). In a similar way, the missing electron (or hole) energy in the HOMO could, although not in all cases, be expressed.

If we assume that electron and hole interact in a screened Coulomb manner we obtain the Hamiltonian for the lowest excited state:

$$H = \frac{-\hbar^2}{2m_e}\nabla_e^2 - \frac{-\hbar^2}{2m_h}\nabla_h^2 - \frac{e^2}{\varepsilon|r_e - r_h|} + \text{polarization terms} \quad (11)$$

and the lowest eigenvalue (i.e. lowest excited state) approximation is:

$$E^* \cong E_g + \frac{\hbar^2\pi^2}{2R^2}\left(\frac{1}{m_e} + \frac{1}{m_h}\right) - \frac{1.8e^2}{\varepsilon R} + \text{smaller terms.} \quad (12)$$

It could easily be seen that E^* is shifted to higher energies due to the second term in Eq. (12). Analyzing Eq. (12), one can say that the energy in confinement is changed to the value of a quantum particle-in-a-box energy of both electron and a hole (term 2) and the shielded Coulomb electron-hole interaction (term 3). The effects caused by polarization of a particle and the correlation in electron-hole mutual motion is included in minor terms of Eqs. (11) and (12). Due to the fact that correlation represented by the third term is not strong, the major effect is additive for electron and hole.

The apparent band gap is thus increased for small R. The predicted approach of E^* to E_g is shown in Fig. 7. The source of such a small approach to bulk properties seems to be caused by a strong chemical bonding. The related question is the dependence of ionization potential on size. This can be modeled by a combination of size dependence of the HOMO (as previously described) with size dependent electrostatic energy of a charged dielectric sphere. The decrease of ionization potential, which could be derived from Eq. (12), opens the possibllity of controlling photochemical processes via particles (see below).

4.1.4 Further Localization of Electrons on Trap States

The further localization takes place if the particle possesses point defects (usually on the surface) which can trap generated electrons and holes. The correlation diagram is depicted in Fig. 8. As can be seen from the picture, the spatial overlap between the hole and the electron is drastically changed if one of them is trapped by the surface state. The presence of a trapped electron-hole pair on the surface trap state determines the different behavior of the exciton: both its life-time and oscillator strength are changed [49] (see Fig. 8). This effect determines the nonlinearity in optical absorption of nanocomposites. The luminescence meas-

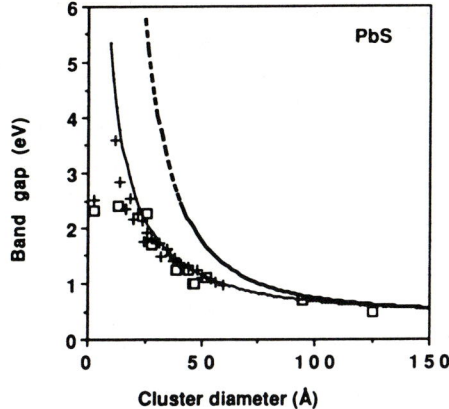

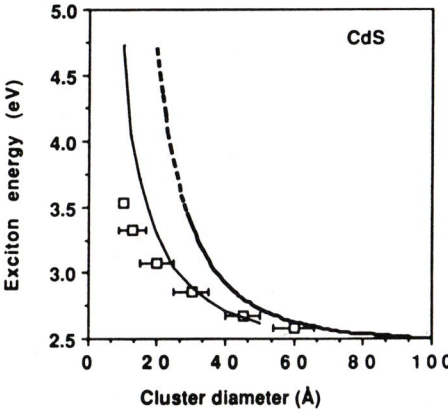

Fig. 7. The dependence of the optical bandgap (exciton energy) of PbS and CdS on the cluster size. *Dashed lines* represent calculation, based on the effective mass approximation. *Squares* represent experimental data. In PbS, *the crosses* are results of cluster tight-binding calculations and *the solid line* is from hyperbolic band modelling. In CdS, *the solid line* is the result of tight-binding calculation [13]

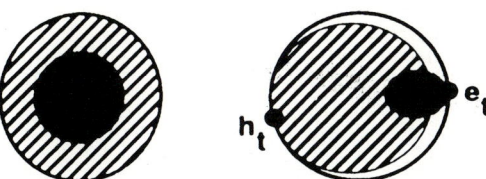

Fig. 8. Hole wave function (smaller, darker circle and electron wave function in condition of absence (*left*) and presence (*right*) of trapped electron-hole pair [13]

urements give evidence to the existence of trapped electron-hole pairs, usually of one pair per small particle [12]. The absorption spectra, however, are dominated by peaks corresponding to thresholds between the states determined by Eqs. (7) and (12).

4.2 Optical Properties of Nanoparticles

The classical approach to the optical spectra of small particles is presented by Mie scattering theory, which postulates that a nanocrystallite is characterized by the same complex $\varepsilon(\lambda)$ as the bulk material. In the relevant theoretical equation (derived on the assumption that the crystalline dimensions are much less than the optical wavelength λ), only the electric dipole term needs to be retained and the extinction cross section σ reduces to:

$$\sigma = \frac{8\pi^2 a^3}{\lambda} \operatorname{Im}\left(\frac{\varepsilon' - 1}{\varepsilon' + 2}\right), \tag{13}$$

where a is sphere radius and ε' is ratio of the complex dielectric coefficient of the bulk crystal $\varepsilon(\lambda)$ to the real dielectric coefficient of matrix. λ is the wavelength in external medium. Eq. (13) was not experimentally verified. The dielectric coefficient must effectively be a function of the crystallite size.

4.2.1 Optical Absorption

Optical absorption spectra of small particles are determined by HOMO → LUMO transition, with the oscillator strength close to unity (Fig. 6). The broadened exciton lines dominate the tail of the spectrum. The measure of crystallite size at which these effects begin to become important is given by the diameter of 1S exciton in bulk crystalline material [12]. For CdS this diameter is equal to ~ 60 Å.

Table 1 shows the effective band gap as a function of crystallite size for CdS crystallites. The experimentally observed spectra (Fig. 9) are then the sum of widely differing spectra corresponding to various sizes, i.e. inhomogeneously

Table 1. Calculated absorption thresholds ("effective band gaps") for CdS spherical crystallites of the indicated diameters. The shift is to higher energy than the bulk band gap. For calculations the CdS band gap has been taken to be 2.53 eV. Calculations are from the elementary model, which overestimates the shifts for the smaller diameters [7]

Diameter (Å)	Shift (eV)	Threshold λ (NM)
Bulk	0	490
75	0.07	478
66	0.10	471
57	0.16	461
48	0.26	444
39	0.45	416
30	0.83	367
21	1.8	286

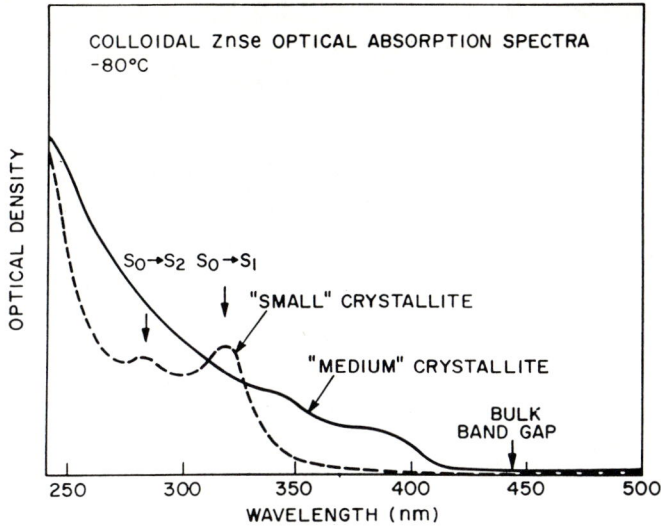

Fig. 9. Size-dependent electronic adsorption spectra of colloidal ZnSe clusters [12]

broadened. It was predicted, however, that the absorption strength at the threshold must be size-independent [50]. That is, it scales with the number of crystallites of a given size. In this case the model uses crystallite size distribution to synthesize the experimentally observed absorption threshold spectrum.

4.2.2 Luminescence Spectra

The tail of the spectra observed could refer to trapped carriers which play the main role in luminescence spectra (Fig. 10). Considering the luminescence phenomenon, the major difference arises between metal and semiconductor particles. The transition metals have an extremely high density of electronic states above the HOMO. The creation of excited states is followed in this case by ultrafast electronic to vibrational radiationless transition. There are no reports of long-lived excited states of luminescence [12]. Semiconductor nanoparticles typically have $S_1 \rightarrow S_0$ energy gap of several electron volts, as has been shown [12] neglecting the presence of trap surface states. Direct $S_1 \rightarrow S_0$ internal conversion is a very high order process in lattice phonons ($h\omega \sim 199\text{--}300 \text{ cm}^{-1}$) and should be slow (Fig. 10).

But as the ratio of surface to volume is high, the surface traps plays an important role in optical properties of small particles. The effects of a surface trapped electron-hole pair on the optical absorption spectrum of CdS particles has been studied with time-resolved laser spectroscopic techniques [50]. The excited exciton is rapidly trapped by surface to form a trapped electron-hole pair,

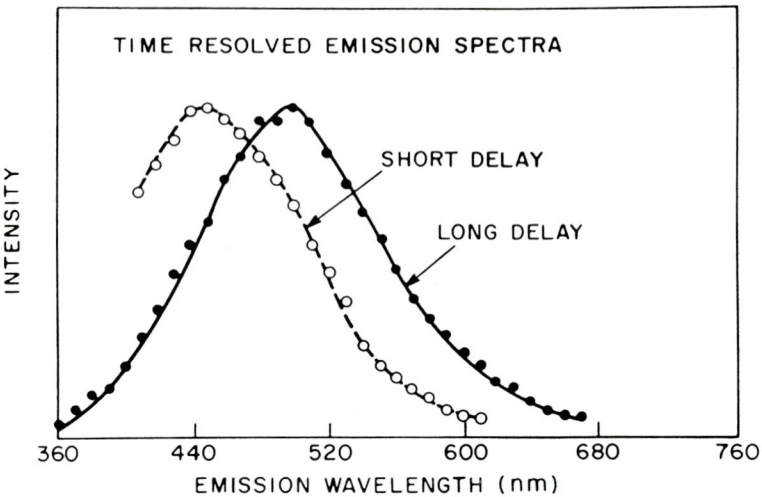

Fig. 10. Time resolved emission spectra of luminescence from ~ 22 Å CdS clusters in a frozen organic glass at = 10 K. Short delay refers to the period 0–1.5 10^{-6} s, long delay to the period $(17-34) \times 10^{-6}$ s. There is no resolvable vibronic structure at = 1 Å resolution under the present conditions [12]

which subsequently decays away with complex kinetics over a time ranging from tens of picoseconds to nanoseconds.

The oscillator strength of an exciton is given as:

$$f = \frac{2m}{\hbar^2} \Delta E |\mu|^2 |U(0)|^2 \qquad (14)$$

where m is the electron mass, ΔE is the transition energy, μ is transition dipole moment, and $|U(0)|^2$ represents the probability of finding electron and hole on the same site (the overlap factor). Assuming that the exciton moves from site to site coherently and is delocalized throughout the whole cluster (similar to delocalization of Frenkel exciton in molecular cluster), the discrete energy band of an exciton can be constructed. At the low-temperature limit the oscillator strength of all levels are concentrated on the lowest exciton state. Thus f per cluster increases linearly with the cluster volume at the low-temperature limit. This is the so-called "giant oscillator strength" effect [51, 52].

4.2.3 Giant Dipole Resonance in Nanoparticles

In nano-size metal particles or thin band semiconductors the above mentioned effect of giant oscillator strength is much stronger due to a large number of electrons on the Fermi level and the possibility of collective excitation of Migdal type [53]. This effect can be interpreted as a coherent vibration of all electrons or holes on definite degenerate energy level from one side of confinement potential to another. In this case the frequency of this vibration is given simply by $f \sim V_F/R$, where V_F – velocity of carrier near the Fermi level and R – radius of small particle. This causes the red shift of plasma frequency in metal particles of small sizes [54]. The red shift observed in this case is interpreted in [55] as being caused by such collective excitation in a confinement potential, the effect similar to giant dipole resonance in atomic nuclei [53].

It has been deduced that the effective cross-section in the interval of giant dipole resonance is expressed by equation

$$\sigma(\omega) = \frac{2\pi^2 e^2 N}{mc} \delta(w - w_m), \tag{15}$$

where ω – frequency of light, N – number of valence electrons, c – light velocity, and

$$\omega_m^2 = \frac{10 E_F}{3 m R^2},$$

where R – particle radius. Another collective excitation in the situation of crossing of the energy levels of electron spectrum is the mode similar to a surface plasmon. The excitation resonance frequency in this case,

$$w_p = \frac{\pi}{3\sqrt{2}} \sqrt{\frac{4\pi e^2 n}{m}}, \tag{16}$$

practically coincides with the value for surface plasmon frequency. The expression for cross-section in this case is:

$$\sigma_p(\omega) = \frac{2\pi e^2 N}{mc} \frac{\Gamma_p}{(\omega - \omega_p)^2 + \Gamma_p^2}. \tag{17}$$

Here Γ_p is the width of plasmon degradation. The condition $\omega_p \gg \omega_0$, (where ω_0 is the quantum transition to the first excited level) is valid for the particles radii $R > 10^{-7}$ m. The additional red-shifted peak caused by w_m was frequently observed [54–56].

4.3 Catalytic Properties of Nanoparticles

The property of nanoparticles which is of most importance to chemists is their special catalytic activity. The small particles not only appear to possess catalytic properties which are not observed in bulk materials, but they also exhibit the specificity of reactions, and the selectivity of catalysts is changed with the decrease of particles size to nanometer range. First of all, it is evident that by varying the particle size we can vary the electron affinity of small particles. Simple consideration shows that the ionization potential is size dependent, thus allowing us to control the reaction ability in ox-red reactions [12]. Another aspect is that the band gap of small particles, and hence the absorbance interval, can be controlled by size. Both these phenomena lead to the increase of the redox potential of photogenerated carriers in small particles. Reduction reactions that cannot occur in bulk materials can thus occur in sufficiently small particles. In [57] a variety of redox couples in aqueous solutions were used to cause the charge transfer between colloid semiconductor particle and redox couple $[R^{n+}/R^{(n-1)+}]$:

$$R^{(n+1)+} + \text{colloid} \rightarrow R^{n+} + (e^-) \text{ colloid} \tag{II}$$

for PbSe and HgSe 50 Å colloids. The position of the lowest vacant electronic state was determined by observing which redox couple could inject electrons into the semiconductor particles. Important examples of photoreduction processes that could only be achieved with small particles of certain semiconductors are the H_2 evolution with aqueous colloids of HgSe (D < 50 Å) and PbSe (D < 50 Å) and CO_2 reduction to formic acid from CO_2-saturated aqueous solutions with CdSe colloids (D < 50 Å). In the former two cases the initial quantum efficiency is about 1–2% in the presence of EDTA (pH 6) or S^{2-} (pH 8.5) as hole scavengers. Another example of the photochemical process is the water reduction by small (~ 40 Å) CdS ruthenium-coated particles synthesized in negatively charged bihexadecyl phosphate vesicles in the presence of thiophenol (PhSH) in aqueous solution [44]. In this case the photogenerated electron is trapped by a ruthenium surface atom, which then undergoes reaction with a H_2O molecule to form hydrogen. The problem that must be overcome is the enhanced photodissociation of the crystallite itself if these clusters are to be used as stable photosensitizers [58]. Photodissociation becomes more pronounced in smaller clusters as excited-state energies increase and the binding energies per atom decrease.

The other way of utilization of catalytic properties in small particles involves the extremely high free surface of small particle catalyst systems. A number of nano-size composites prepared from noble metals exhibited enhanced catalytic activity and long-time stability. The Au/Pd bimetallic cluster catalysts in poly-(N-vinyl-2-pyrrolidone), of size ~ 20 Å, were synthesized by successful reduction of $HAuCl_4$ and $PdCl_2$ solutions [20].

The main reactions are:

Nucleus Formation:

$$2 Au^{3+} + 3 C_2H_5OH \rightarrow 2 Au + 3 CH_3CHO + 6H^+ \tag{III}$$

$$Pd^{2+} + C_2H_5OH \rightarrow Pd + CH_3CHO + 2H^+ \tag{IV}$$

Growing process

$$Pd_m + Pd^{2+} + C_2H_5OH \rightarrow Pd_{m+1} + CH_3CHO + 2H^+ \tag{V}$$

$$2 Au_n + 2 Au^{3+} + 3 C_2H_5OH \rightarrow 2 Au_{n+1} + 3 CH_3CHO + 6H^+ \tag{VI}$$

$$2 Pd_m + 2 Au^{3+} + 3 C_2H_5OH \rightarrow 2(Pd_mAu) + 3 CH_3CHO + 6H^+ \tag{VII}$$

$$Au_n + Pd^{2+} + C_2H_5OH \rightarrow Au_nPd + CH_3CHO + 2H^+. \tag{VIII}$$

The EXAFS studies showed the particles to have a cluster in cluster structure. Depending on the concentration of chlorides, the clusters were Au-wrapped Pd clusters or Pd-wrapped Au structures. The latter exhibited enhanced catalytic activity in the reaction of hydrogenation of 1,3-cyclooctadiene in ethanol at 30 °C under 1 atmosphere of H_2. The catalytic activity increased with decreasing particle size.

4.4 Debye Length Limit in Nanoparticles

The last distinctive feature of small semiconductor particles is the partial screening of Coulomb interaction which takes place in semiconductor particles if the particle radius is less than the Debye length for semiconductor:

$$L_D = (\varepsilon \varepsilon_o kT/e^2 n_B)^{1/2} \tag{18}$$

where ε – dielectric coefficient of semiconductor crystallite, n_B – number of free charge carriers in conduction band of a particle. This fact is caused simply because, if the number of free electrons is insufficient, then the external field is not screened and hence field penetrates the particle, changing the position of Fermi level. This fact plays an especially important role for catalysts (see above) and semiconductor chemical sensors due to the fact that, in this case, surface charge caused by chemisorption is not screened and causes the drastic change in concentration of charge carriers due to the shift of Fermi energy level.

5 Properties of Polymer Nanocomposites (Cooperative Properties of Systems of Nanoparticles)

In the previous section the properties of nanoparticles themselves were considered; such properties were demonstrated by various systems containing nanoparticles [3–11]. This section is devoted to the description of the cooperative properties of nanosize particles in polymer matrix where the nature of a matrix is important. A number of anomalous properties exhibited by such composites consists of the interaction of small particles through the polymer matrix. In such processes the properties of the polymer matrix cannot be neglected and influence the cooperative behavior of the systems.

5.1 Conductivity in Polymer Nanocomposites

5.1.1 Interparticle Charge Transfer

Associated with the problem of conductivity of polymer composites filled with conductive particles, there are the following three different ranges of conductivity depending on filler concentration.

1. The range of low filler concentrations, where conductivity is determined by polymer matrix conductivity by a mechanism of hopping through impurity centers. In this range a filler particle can only inject some carriers into the polymer matrix, but does not affect the conductivity process to a great extent.

2. The range of high filler concentrations, where the conductivity is determined by charge transfer through the system of contacting filler particles or "Infinite Cluster". This conductivity can be of non-activation type if the interparticle barriers, similar to those in polycrystalline ceramics, are lower than the Fermi level in a particle. In the opposite case, there exists activation energy determined by the height of these interparticle barriers.

3. The third range is the threshold zone where the conduction mechanism is changed from the first to the second. Such a threshold is similar to the Metal-Insulator threshold which has been intensively studied by condensed matter physicists [62]. In the case of polymer-composites this threshold can be determined as a percolation-type of metal-dielectric threshold due to the fact that the change of conductivity mechanism takes place when the filler particles from the conductive cluster spanning the entire volume of composite. Since the distribution of particles in the matrix is random, the threshold is determined by geometrical and topological characteristics of the medium and described by the percolation theory, which is the part of probability theory dealing with such problems of geometrical and topological origin [63].

The investigations of polymer-nanocomposites by Godovski [2] and Godovski and Sukharev [61] have shown that the metal-insulator threshold for nanocomposites is drastically different from the one for composites with larger particles. Let us consider the situation with closer nanocomposites.

Regarding the question of interparticle charge transfer between nanoparticles through the polymer matrix, two main mechanisms seen to dominate: the hopping of electron or holes either directly or consecutively through hopping centers in polymer matrix and the process of injection of charge carrier in matrix conduction band either by thermoelectronic emission or electric field emission and the charge transfer to another particle or through matrix. For low concentrations of small particles the second mechanism seems to dominate. The activation energy of carrier injection is usually high (2–4 eV) which makes the process probability not very high. Thus, the conductivity in this case hardly exceeds the conductivity of a polymer matrix itself. However, it can be increased by photoinjection or in strong electric fields, which decreases the activation barrier height as predicted by the Pool-Frenkel law [5]:

$$I = \text{const } NT^m E^n \exp\{e\beta E^{1/2} kT\} \exp\{-\varphi_i/kT\} \tag{19}$$

where E-electric field, φ_i-height of the barrier. The hopping probability decreases with the increase of interparticle distance according to the expression [59]:

$$P_{ij} = P_{ij}^o \exp\{-2r_{ij}/a - 2q^{3/2} E_{ij}/\varepsilon^{1/2} kT - q^2/\varepsilon RkT\} \tag{20}$$

which is obtained regarding the spatial overlap of the exponential tails of wave functions in neighboring particles, and assuming the potentials between particles to be mostly of Coulomb origin:

$$I_{ij} = \int \frac{\psi_i(r)\psi_j(r)e^2}{\varepsilon(r_i - r_j)} - \psi_i(r)\psi_j(r)dr' \int \frac{e^2\psi_i(r)dr}{\varepsilon(r_i - r_j)} \quad (21)$$

where ε-dielectric coefficient of the medium, ψ_i, ψ_j-wavefunctions of i- and j-eigenstates in neighboring particles, determined in the simplest case by the expression:

$$\psi(r) = \exp\left\{\frac{me^2r}{\hbar\varepsilon}\right\}. \quad (22)$$

The activation energy in this case [5] is the energy of carrier generation, mostly of electrostatic origin. It refers to a process of electron-hole pair generation.

The hole and electron are generated on the neighboring particles, the following process of carrier transfer being assumed to be isoenergetic. The various models developed on the basis of Eq. (20) take into account the presence of other neighboring particles, the possibility of transfer through hopping centers in polymer matrix, and so on. The diagram exhibiting the various processes of transfer is shown in Fig. 11. The multiple hopping through the hopping centers formed by impurities in the polymer matrix or defects in the polymer chain could increase the hopping distances to hundreds of angstroms [60]. The final possibility of the transfer is the multiplied probability of single steps:

$$Ps = \prod_{i=1}^{n} P_i,$$

where P_i is determined as in Eq. (20).

5.1.2 Peculiarities in Percolation Behavior of Electroconductivity

The observation of electrophysical properties and, primarily, the electroconductivity of nanocomposites, shows a number of differences from large-size

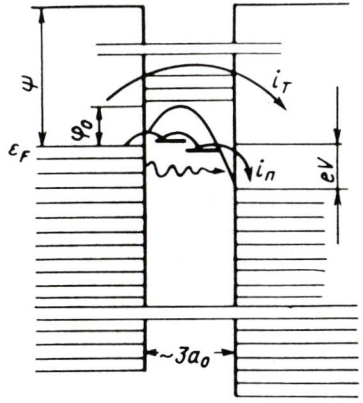

Fig. 11. Energetic diagram of the interface metal–dielectric (polymer)-metal-work function of extortion of electron to vacuum. ψ_o-work function of extortion from metal to dielectric, E_f-Fermy energy of the system. V–voltage on the contact, a_o-constant in expression $i_T = a_o V \exp\{-e_o/kT\}$, where i_T-tunnel current of interfacial transfer. electrostatic energy, i_n-current of dielectric–metal transfer, i_T-current, caused by thermal excitations in metal

systems. In [61] the analysis of such differences has been conducted for the CuS-polyvinyl alcohol composites with CuS microcrystals of size 15–20 nm which showed the anomaly in percolation behavior of electroconductivity of nanocomposites (Fig. 12). The main difference observed was in the large-size filled composites (curves 1 and 2 respectively).

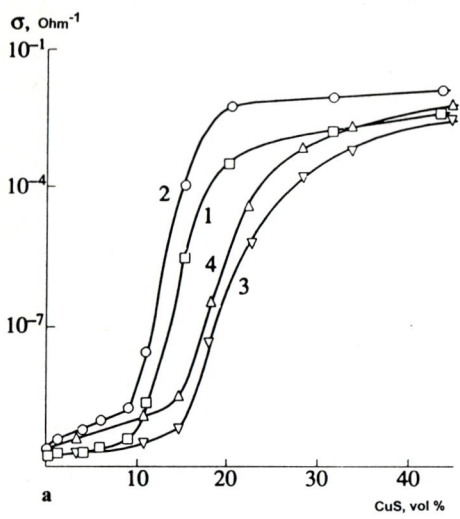

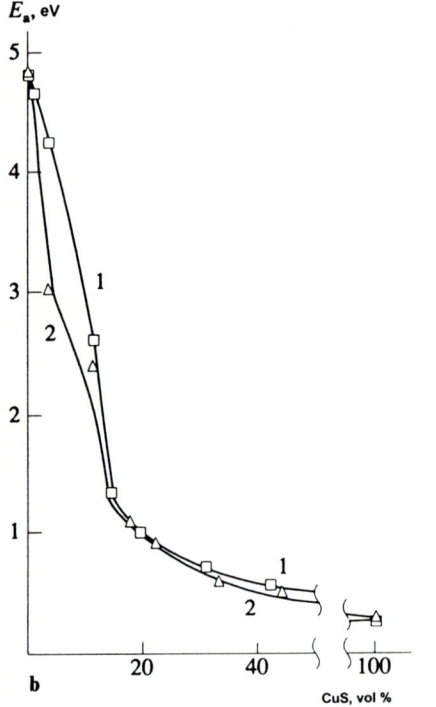

Fig. 12a. The dependence of conductivity on filler concentration for nanosize 120 Å CuS-PVA composites (*curves* 1,2) and 10 μm CuS-PVA (*curves* 3,4) composites. b Activation energy concentration dependence for 120 Å CuS-PVA and 10 μm CuS-PVA [2]

The difference in the behavior of magnetoresistance (Fig. 13) was shown by large positive values of magnetoresistances reached before the percolation threshold. The dependence of magnetoresistance on magnetic field in the area of such positive values of magnetoresistance could be linearized in coordinates Ln $R_H/R_o - H^2$ (I) (Fig. 13b). The latter fact, together with the Mott character of thermal dependence of electroconductivity (Fig. 14): $Ln\sigma \sim 1/T^{1/2}$ (II), has been interpreted as the existence of hopping conductivity in polymer composites. The dependence of conductivity on filler concentration, which could be described as $Ln\sigma \sim 1/X_v^{1/3}$ (III), points to the fact that the centers of hopping in this case are nanoparticles themselves.

It was assumed that hopping consists primarily in tunneling of charge carriers (holes) between microcrystals through polymer layers which could contain virtual states adding to the amplitude of direct process the amplitudes of non-direct transition. The equation for conductivity in that case is:

$$R_{ij} = R_{ij}^o \exp\{2r_{ij}/\beta + W_{ij}\}, \tag{23}$$

where r_{ij}–intergrain distance, W_{ij}–activation energy of $i \to j$ transition, β–the characteristic radius of eigenstate (for the confinement potential it is determined by expression:

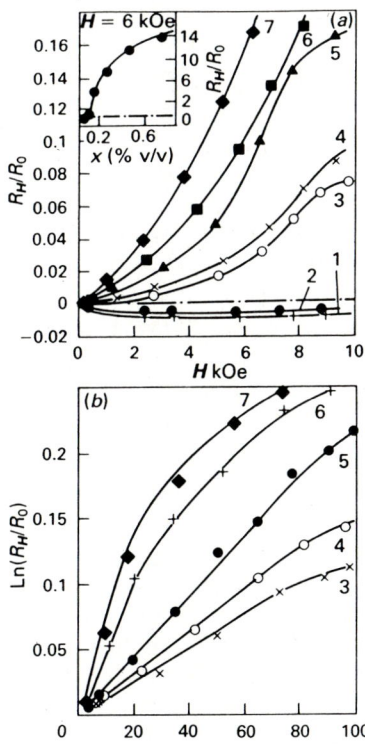

Fig. 13a. Dependence of CuS-PVA 120 Å composite magnetoresistance on magnetic field. On insertion–the dependence of magnetoresistance at magnetic field $E = 6$ kOe on CuS concentration in composition. b Dependence of magnetoresistance (R(H)/R(O)) in coordinates Ln R(H)/R(O)) – $H^2(Oe)^2$ [61]

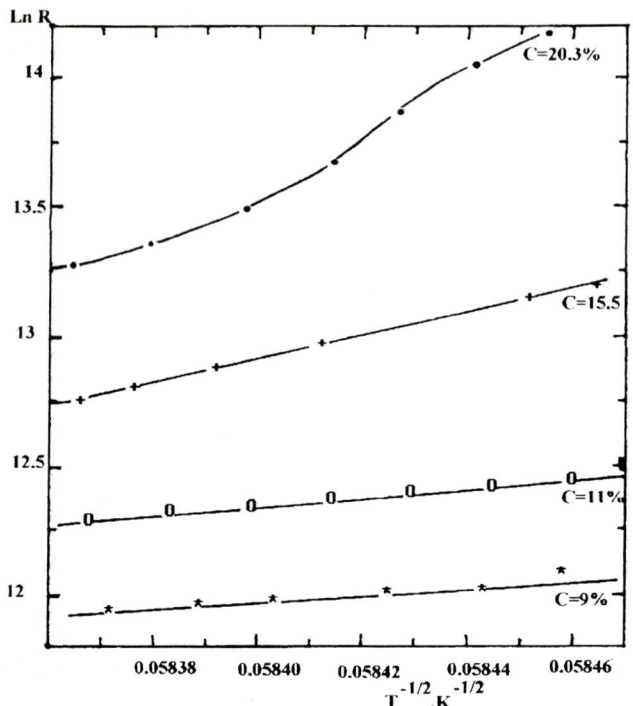

Fig. 14. Dependence of logarithm of CuS-PVA 120 Å composite resistance on $1/T$, $K^{-1/2}$ [61]

$\beta = h\sqrt{2mE_o}$, where E_o–the energy of level and finally $R_{ij}^o = kT/e^2 v_o$, where v_o the frequency factor). The main assumption is that the formation of a true infinite cluster of contacting particles is preceded by appearance of the cluster with hopping conduction between microcrystals of the filler. Thus we have the specific effect of nanocomposites – the dominance of hopping conductivity in the system for the concentrations preceding the formation of the "true" infinite cluster, consisting of contacting nanoparticles. Theoretical consideration of possible conductivity mechanisms in such systems points out two possible mechanisms. For the relatively small amount of filler particles the Shklovsky-Epfros [59] approach could be applied, developed for doped compensated semiconductors on the basis of standard percolation theory. The expression for conductivity in this case is:

$$\sigma = \sigma_o \exp\{-\alpha/\beta N^{1/3} - q^2/\varepsilon kT + 2(q^3 \bar{W}/\varepsilon(kT)^2\}^{1/2} \qquad (24)$$

where $\alpha = 1.73 \pm 0.03$, N–concentration of particles in the volume unit, which is connected with volume fraction of filler by the expression $N = (3/4)X/\pi R^3$. The final expression obtained is:

$$\sigma = \sigma_o \exp\{-(4\pi/3)^{1/3} \alpha R/X^{1/3} - q^2/\varepsilon kT + 2(q^3 \bar{W}/\varepsilon(kT)^2\} \qquad (25)$$

which describes a number of experimental facts (the dependencies (I–III, p. 27)). For the higher concentrations of filler, when the optimization on the energy of hopping becomes possible, the Mott [62] approach could be applied and the expression like:

$$\sigma = \text{const} \exp\{-A/V\sqrt{kT\varepsilon}\} \tag{26}$$

(which is obtained by choosing the optimal particle diameter as that which minimizes the exponent in Eq. (23)) is derived.

The existence of hopping conductivity in the system leads to the discrepancies from "classic" percolation behavior in the systems, which consist of a shift of X_c–the filler content of the onset of infinite cluster, which is 0.17 + 0.01, as predicted for solid non-interpenetrating spheres problem by percolation theory. Another difference from the classical spheres problem is the discrepancy between conductivity critical index t in the expression [63]: $\sigma = \sigma_o(X - X_c)^t$ and the value given by the classic spheres problem (2.4 ± 0.1 in contrast to 1.9 ± 0.1) (Fig. 12). Both these facts could be described in terms of a simple model, proposed in [59], which assumes the existence of delocalization of charge extending the nanoparticle. Thus the situation could be described by a two-spheres model (see Fig. 15.), one of them depicting the particle and the other sharing the characteristic radius of charge depletion in polymer matrix. The percolation in terms of this model begins when external spheres intersect, and thus percolation arises before direct contact between grains. The characteristic parameter R_d/R_g determines the value of shift. Monte-Carlo computer

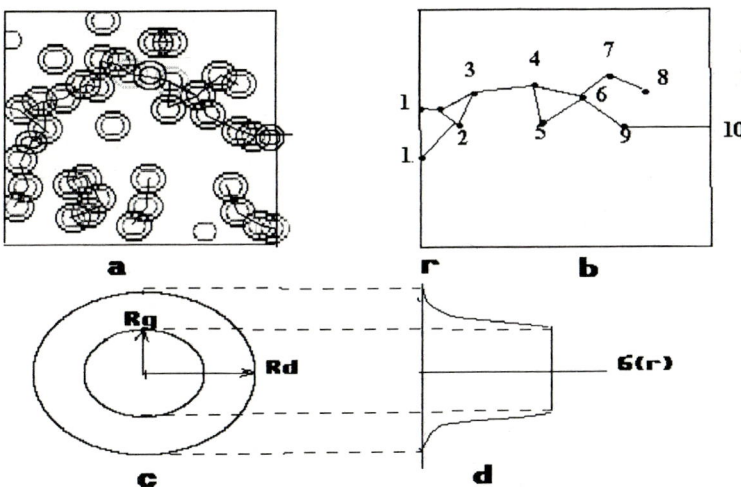

Fig. 15a. Distribution of 2-dimensional spheres with spatial restrictions on the plain (the percolation from left to the right side is reached. **b** Equivalent graph of the random resistors net, corresponding to case A. **c** Structural unit of the net, depicted in **d**. The distribution of conductivity over the unit length

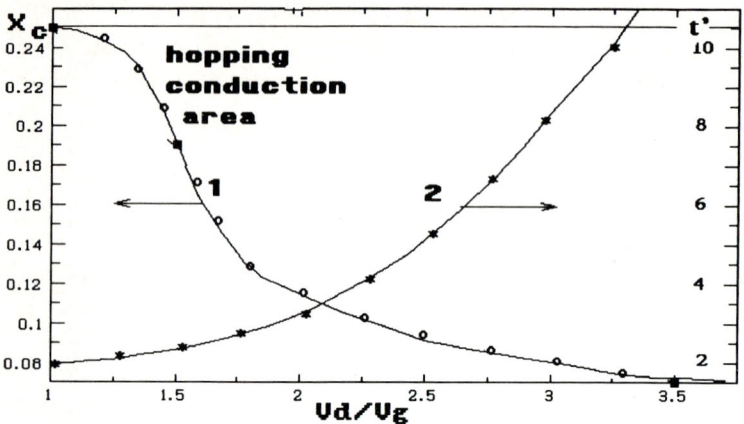

Fig. 16. The dependence of percolation threshold value X_c (vol. %), curve 1, and the value of conductivity critical index t (see text) on parameter Vd/Vg. (Vd is the volume of internal sphere (Fig. 15), Vg–the volume of external sphere)

simulation of the dependence of X_c on R_d/R_g is depicted in Fig. 16 (region 1) and is in good agreement with the experimental data. The computation of random resistor network conductivity shows that conductivity critical index t must also depend on the value of R_d/R_g. Figure 16 (region 2) shows this dependence for three-dimensional cases.

5.1.3 The Influence of Polymer Matrix on Charge Transfer

The influence of polymer matrix on the behavior of electroconductivity found for this nanocomposite system was the drastic increase of composite conductivity caused by absorption of water vapors by polymer matrix. The value of such a change has the sharp extremum corresponding to the range where hopping conduction dominates the system electroconductivity. It was assumed that the influence is primarily realized through the dependence of hopping conductivity (see Eq. (25)) on dielectric constant of polymer matrix, which is drastically changed when absorption of such polar molecules as H_2O takes place. The simple theoretical model has been developed (mean field approximation):

$$\bar{\varepsilon}_{mix} = \bar{\varepsilon} - \overline{(\delta\varepsilon)^2}/3\bar{\varepsilon}, \tag{27}$$

where ε_{mix}–dielectric constant of mixture, $\bar{\varepsilon}$–volume mean dielectric constant. One may derive from (Eq. 27), neglecting the second term,:

$$\varepsilon \cong \bar{\varepsilon} = \varepsilon_{H_2O} y + \varepsilon_{PVA}(1-y) \tag{28}$$

where ε_{H_2O} and ε_{PVA} are the dielectric constants of water and PVA respectively, y-volume concentration of water in matrix.

It was assumed that before formation of infinite clusters with hopping conductivity, the conductivity takes place through finite clusters of nanoparticles, with the distances allowing the hopping transfer. Such finite clusters are separated by large distances which cannot be covered by hopping; thus the carrier transfer takes place by the motion in the conductivity band of the matrix. The expression for hopping conductivity can be derived from the condition of dominance of hopping conductivity, taken from Eq. (23) over the polymer matrix band conductivity:

$$r = \beta/2 \left[\operatorname{Ln} \frac{qv_o}{kT\sigma_m} - \frac{\bar{W}}{kT} \right] \tag{29}$$

where σ_m–ohmic type conductivity of polymer matrix.

Thus percolation theory gives the following expression for conductivity before formation of an infinite cluster with hopping conductivity (the problem of grey-black percolation) [65]:

$$\sigma = \sigma_m (p_H - p)^{-q} \tag{30}$$

where q–the critical index of conductivity through finite clusters, in this case $q \cong 1$, $P = 4\pi N r^3/3$–parameter, determining the mean number of bonds per one particle, $p_H = 2{,}7$–critical value of bonds number. Substituting the expressions for p_H and p, the following equation for conductivity in this range can be obtained:

$$\sigma = \sigma_m \left(2.7 - \frac{\pi}{6} N \left(\varepsilon \frac{\hbar^2 \varepsilon_o}{me^2} \operatorname{Ln}(e^2 v / kT\sigma_m) - \hbar^2 / kTm\bar{R} \right)^3 \right)^{-1}. \tag{31}$$

Assuming that conductivity response for water absorption is determined primarily by change of hopping conductivity in finite clusters, we obtain:

$$s = \frac{\sigma - \sigma_o}{\sigma_o} = \frac{2.7 \frac{\pi}{6} N (A\varepsilon_i - B)^3}{2.7 \frac{\pi}{6} N (A\varepsilon - B)^3} - 1 \tag{32}$$

for the relative change of conductivity (sensitivity) under water absorption. Here $A = (\hbar^2 \varepsilon_o/me^2)\operatorname{Ln}(e^2 v_o/kTm)$, $B = \hbar^2/kTm\bar{R}$, ε_i–initial value of dielectric constant before water absorption for $X < X_h$, which corresponds to the range, where formation of infinite cluster with hopping conductivity takes place.

For the range of filler concentrations above that, corresponding to formation of infinite cluster with hopping conductivity and before the true percolation in the system, the following expression for conductivity, obtained for conductivity through infinite cluster with correlation length L_o, has been used [59]:

$$\sigma = \sigma_o \exp\{-\bar{W}/kT\} \exp\{-\zeta_c\}, \tag{33}$$

where $\zeta_c = 2r_c/\beta$, $\sigma_o = (R_o L_o)^{-1}$, $R_o \cong kT/e^2 v_o$, L_o–correlation radius of a critical subnetwork, i.e. infinite cluster corresponding to $\zeta_{ma} = \zeta_c + 1$ [9]:

$$L_o \cong N^{-1/3} (2r_c/\beta)^v. \tag{34}$$

In fact this is just another way to rewrite Eq. (25) in terms of an infinite cluster of percolation theory. In this case for system conductivity we obtain:

$$\sigma \cong R_o^{-1}(N)^{\frac{1+\nu}{3}} \left(\frac{\hbar^2 \varepsilon_o}{1.73\, me^2}\right)^\nu \varepsilon^\nu \exp\left(\frac{-1}{\varepsilon}\left(\frac{e^2}{\bar{R}kT\varepsilon_i} + \frac{1.73}{\hbar\varepsilon_i N^{1/3}}\right)\right), \quad (35)$$

and for sensitivity to water absorption:

$$S = \frac{\sigma - \sigma_o}{\sigma_o} = \left(\frac{\varepsilon}{\varepsilon_i}\right)^\nu \exp\left\{\frac{\varepsilon - \varepsilon_i}{\varepsilon \varepsilon_i}(\alpha + \gamma N^{-1/3})\right\} - 1, \quad (36)$$

where $\alpha = e^2/kT\varepsilon_o \bar{R}$, $\gamma = 1.73\, me^2/\hbar^2 \varepsilon_o$. Equations (18) and (19) are valid for $X_h < X < X_c$, where X_c corresponds to the formation of true infinite cluster in a system. Thus one can see the conductivity response to ε change for $\varepsilon \gg \varepsilon_i$:

$$S \sim \varepsilon^{0.88}. \quad (37)$$

and for $\varepsilon - \varepsilon_i \ll \varepsilon_i$ we obtain following expression:

$$S \cong (\varepsilon/\varepsilon_i)^\nu \exp\left(\frac{\varepsilon - \varepsilon_i}{\varepsilon_i^2}(\alpha + \gamma N^{-1/3})\right) - 1. \quad (38)$$

The model qualitatively describes the experimental data (see Figs. 18 and 19 with Figs. 17 and 12). The question about the influence of other properties of polymer matrix (e.g. crystallinity, liquid-crystal ordering etc.) on interparticle charge transfer is still unresolved.

5.2 Optical Properties and Photoconductivity Behavior of Polymer Nanocomposites

5.2.1 Dielectric Confinement Effect

The first optical effect pointed out by Wang [13], and studied by computational simulations, is so-called dielectric confinement. Dielectric confinement is caused by the difference in refractive indices of a polymer medium (which has lower refractive index) and a semiconductor or metal particle (which usually has higher refractive index). When illuminated by light, the field intensity near, at and inside the particle surface can be enhanced considerably compared to the incident intensity because of the boundary established by the different refractive indices. This local field enhancement effect can have important consequences on photophysical and nonlinear optical properties of such polymer-nanoparticle systems.

The expression for intensity enhancement factor Q_{NF}, which was obtained using classical Lorentz and Onsager models by Messinger et al. [66] is:

$$Q_{nf} = \frac{R}{\pi a}\int_0^{2\phi}\int_0^\phi \bar{E}_s \bar{E}_s^* \sin\theta\, d\theta\, d\phi_{R=a}. \quad (39)$$

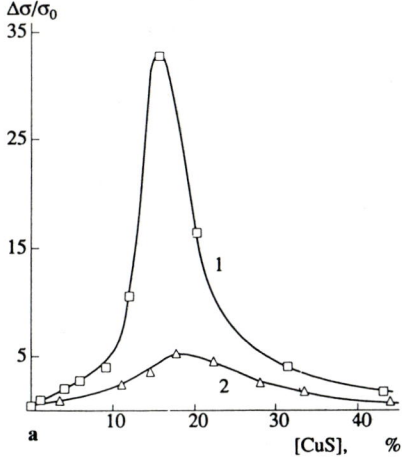

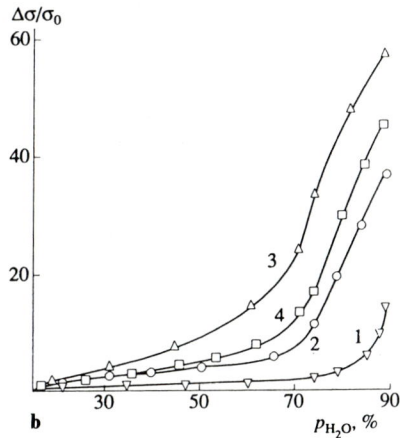

Fig. 17a. Dependence of relative change in CuS-PVA composite conductance on the increase of relative humidity from 15 to 74% on CuS content. *1*–Nanocompositions with particles size ~ 120 Å, *2*–large (10–20 μm) CuS particles compositions. **b** Dependence of relative change of nanocomposites conductivity on relative humidity. Filler concentrations: *1*–7%, *2*–11%, *3*–17.5%, *4*–22% [64]

This equation has been used successfully to calculate local field enhancement in silver particles [66] and water droplets [67]. The application to semiconductor particles in solid media was conducted by Wang [68]. In Fig. 20 one can see the size-dependent local field enhancement factor Q_{NF}, of CdS and GaAs at several selected wavelengths, covering both absorbing and nonabsorbing regions. An interesting application of this phenomenon for optical switching was discussed in [69]. The effects of cluster shape and cluster–cluster interaction on local field effect need to be studied as pointed out in [13].

5.2.2 Non-scattering of Light

One of the anomalous optical properties, which distinguishes the composites with nanoparticles from the systems with large filler size is the possibility of non-scattering of light if the particle size is less than one-fourth of the light

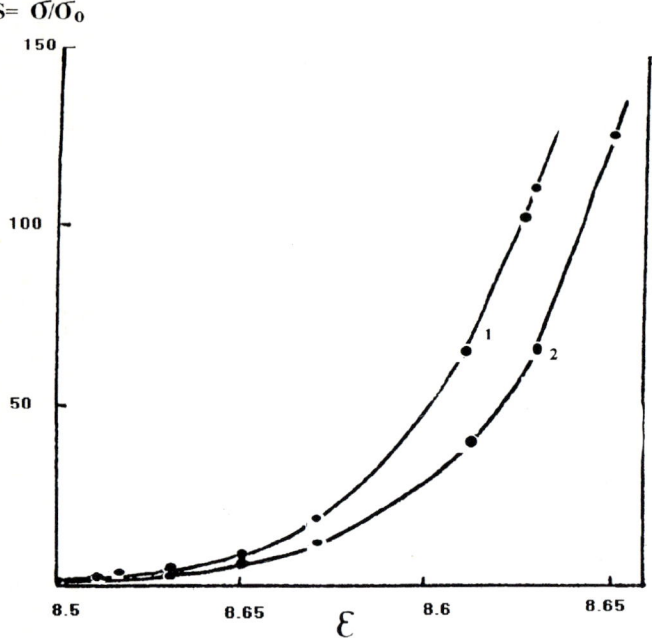

Fig. 18. The simulated dependence of relative conductivity change for CuS-PVA nanosize composite on dielectric constant of (PVA + absorbed water) system [64]

wavelength. For visible light this range is from 2–5 to 50 nm. Thus the combination of polymer matrix with nanocrystals of that size permits one to obtain optically homogeneous media with optical properties similar to those of monocrystals of filler substance combined with elasticity, solubility and good processability of polymer component. The dispersing optical elements, narrow-band and scattering photofilters could be made using these composites.

As an example [35], the transmittance spectrum of thin (10 μm) CdS with particle sizes of 50 nanometers in polyvinylpyridine on a sapphire substratum is shown in Fig. 21. For comparison the transmittance spectrum of a standard glass photofilter GS-11 with a thickness of 5 mm is also given. One could see the coincidence in steepness of the shortwave side of the spectrum, the value of transmittance in the transparency interval from 0.5 to 2.5 μm and the position of the long wave border at about 2.5 μm. Nevertheless, in contrast to the glass, the composite photofilter has one additional transparency interval which is typical for polymer matrix. This band is in fact typical for organic substances and is determined by the vibration of CH-, NH- and OH- atom groups. Composite material with the composition 1:1 by weight has a reflexion index of about 1.7, whereas for the pure polymer matrix it is only 1.4.

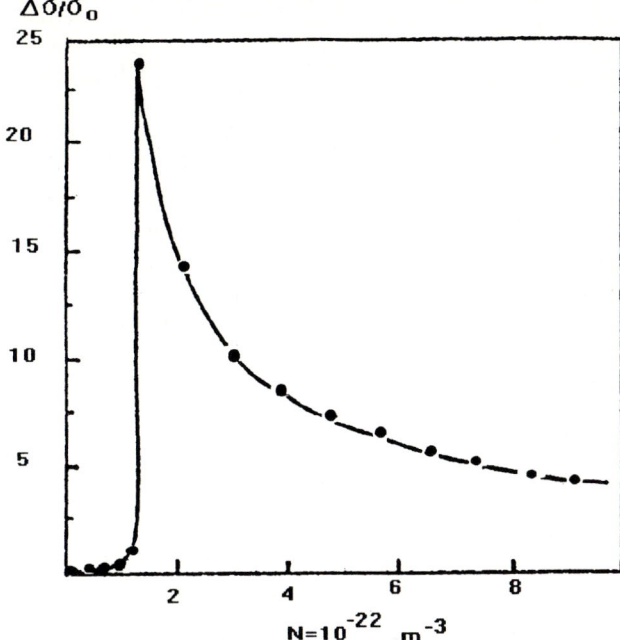

Fig. 19. The simulated dependnece of CuS-PVA nanocomposite conductivity relative change on CuS content in composition [64]

5.2.3 Photoconductivity

Another cooperative effect exhibited by nanocomposites with semiconductor CdS [35] and CdSe [70] particles is the essential photoconductivity which is high, not only for the vales of filler concentration after percolation threshold (see Fig. 22), but even for the compositions with lower concentrations of semiconductor filler. This is determined by good injection of photogenerated carriers in polymer matrix due to a large particle-polymer interfacial area and lower values of work function for nanoparticles [70]. Thus compositions with good transport properties of polymer matrix, such as CdS with polyvinylpyridine with addition of diphenylhydrazine as a molecular hopping component [35] exhibited the homogeneity of polymer substances. The charge mobilities measured in the sandwich cell by photogeneration using a nitrogen LASER LGI-21 with $\lambda = 337$ nm and impulse duration of 10 nsec. The drift mobilities were found to be $\mu_e > 6\cdot10^{-3}$ sm^2/V sec and $\mu_h = 3\cdot10^{-3}$ sm^2/V sec. These values exceed by two orders of magnitude the mobilities for the majority of high-resistive polymer-organic and amorphous semiconductor layers.

Thus such compositions could be applied as low inertial photoreceiving devices in which high resistive semiconductors are applied (e.g., in the method of

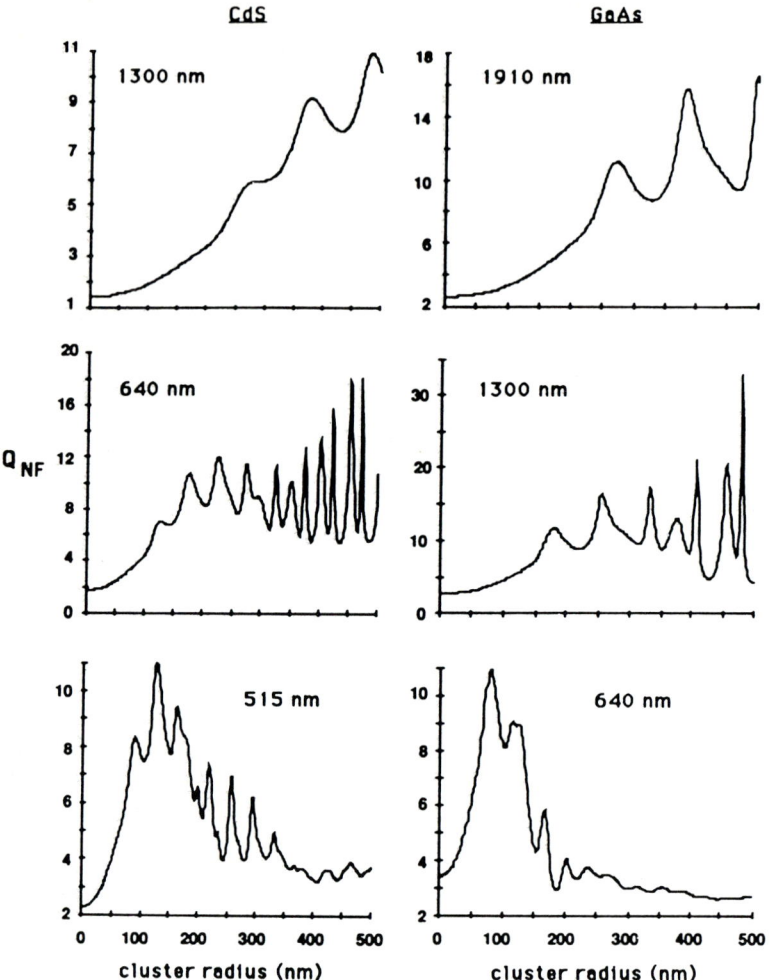

Fig. 20. The size dependent local field enhancement factor Q_{NF} for CdS and GaAs at several selected wavelengths, covering both absorbing and nonabsorbing regions [1]

drift mobility determination in semiconductor materials measurement as the generation layer instead of commonly used amorphous selenium [35]).

The other promising field of application for such composites is the synthesis of materials and devices for electrophotography (EP) and photothermoplasticity (PTP) registration media [35].

The advantages from the application of CdSe, CdS, ZnO and ZnS sensibilized nanocomposites are as follows:

– the combination of high values of photosensitivity, flexibility, optical homogeneity and transparency in the nonsensitivity intervals;

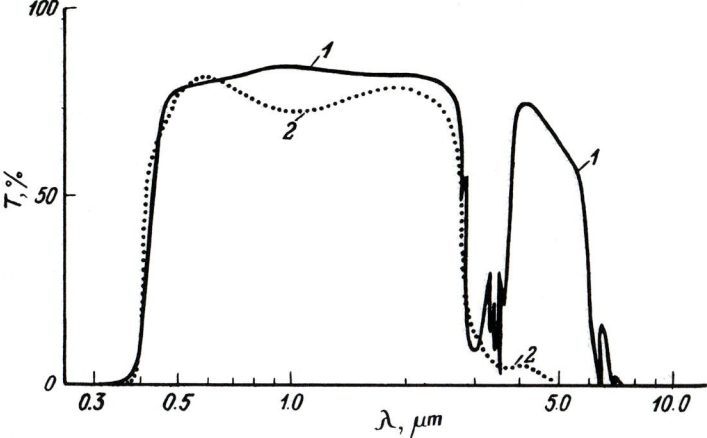

Fig. 21. Transmittance spectrum of composition PVP-CdS nanocrystals (*1*) and glass GS-11 (*2*) [35]

– the ability to apply the broad range of semiconductor filler along with their mixtures for variation of photosensitivity and transparency range of the material;
– the preparation of layers with determined optical density by varying the concentration of semiconductor filler and polymer matrix and also controlling the particles size during synthesis;
– the simplicity of layer preparation– all the stages could be carried out in solution without sputtering and intensive thermal treatment; synthesis of photosensitive composition, its chemical and spectral sensibilization, pouring of the layers, i.e. the traditional technology for the preparation of materials in chemico-photographic production;
– the reproducibility and homogeneity over the entire film length and the possibility of obtaining flexible EP and PTP materials;
– the possibility of synthesizing the composite with varying refractive indices, which can be applied as waveguides.

The samples [35] for the investigation of EP and PTP were synthesized from 50 nm CdS particles in polyvinylpyridine matrix with diphenylhydrazone addition by pouring emulsion on rigid or flexible substratum; the thickness was 3–10 mkm after drying. PTP layers were obtained by creating the TP layer of divinylstyrene on the EP layer with a thickness of 5 mkm.

The layers exhibited the following properties:
The optical homogeneity of EP and PTP layers was comparable to that of organic polymers. Having the thickness of 4–7 mkm, they were electrically charged up to surface potentials of 400–450 V independent of the sign of the voltage on coronator.
The relaxation curves for both interfaces were similar, the residual potential was small (not greater than 1–5 V), the time of dark semidownfall of surface potential

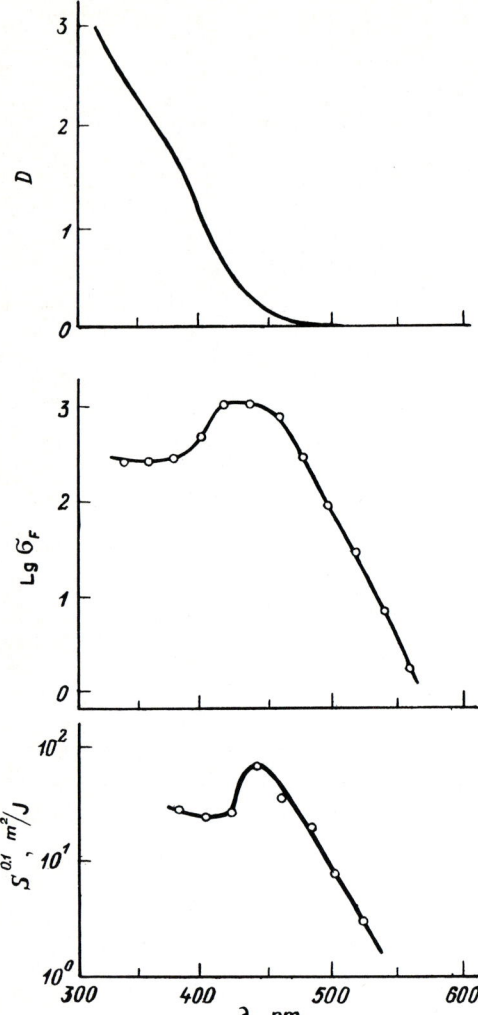

Fig. 22. Spectra of optical density D, photoconductivity, s_a and electrophotographic sensitivity $S^{0.1}$ of layers of nanocompositions of CdS in PVP (the concentration 1:1 by weight) [35]

was about 10^3 seconds (Fig. 23). Such rates of potential downfall in the darkness are acceptable for application to EP devices. The spectrum of EP sensitivity is shown in Fig. 22 (curve 3): it looks very similar to the photoconductivity spectrum (curve 2). The photosensitivity in absorbance band maximum and on its side has the values 100 and 10 m²/J respectively for the surface potential downfall criterion of 0.1 for wavelengths 450 and 500 nm. These values are close to the parameters of existing EP layers and exceed those demonstrated by layers of organic polymers. However the spectral sensitivity of CdS nanocrystals is in the blue range of visible light. For broadening to the long wave area it seems feasible to apply the semiconductors with narrower band gap (as, for example, CdSe).

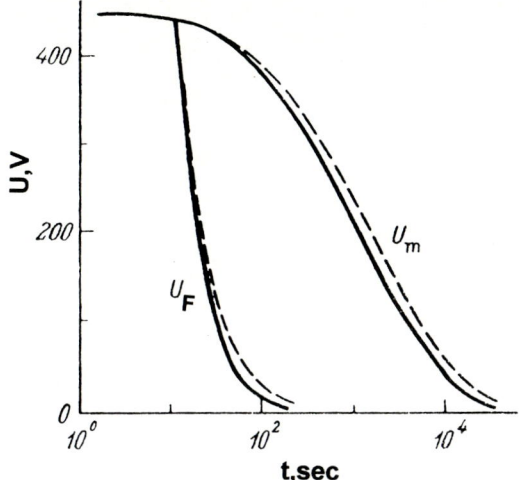

Fig. 23. Relaxation of surface potential of nanocomposition EP layer with the thickness 7 mkm in dark (UT) and under radiation U_a for positive (*dense lines*) and negative (*dashed lines*) voltage on coronator [35]

The double-layer samples, obtained by preparation on composite EP layer of the PTP layer [35] exhibited relatively moderate PTP sensitivity (5 m^2/J) and the criterion of diffraction effectiveness of 1% for 450 nm wavelength. However, the layers were not optimal for reaching the maximal PTP sensitivity and the authors hope to increase the PTP properties of such double layers.

5.2.4 Nonlinear Effects

The other important optical property of polymer nanocomposites is their optical nonlinearity [13, 71]. The compositions of CdS in Nafion were obtained by an in-situ method [71] consisting of ion-exchange of Nafion thin film with aqueous solution of cadmium acetate followed by washing, drying and exposure to hydrogen sulfide. Compositions obtained contained CdS particles of sizes in the range from 30 to 200 Å depending on the moisture content during reaction with H_2S. The increase of moisture content increased the particles sizes.

Degenerate-four-wave mixing, DFWM, with pumping by dye laser of intermediate power, was applied to study the optical nonlinearity of the composite. The essential nonlinearity has been observed in long pulse (nanoseconds) conditions. Both the absorption and the degenerate four wave mixing signal saturate at moderate power (a few MW/cm^2). The experimental data are qualitatively consistent with a three level saturation absorber model [72]. Absence of electron–hole recombination emission indicates a very short lifetime for electron–hole pairs (level 2 of three level model), and rapid trapping by the defect state (level 3 of three level model). It has been pointed out that by changing the surface chemistry one can change the quantity of trapping states, thus affecting the optical properties of the polymer-composite.

In the absence of a transient effect, which applies in the present case since the laser pulse-width is long compared to the defect state lifetime, the DFWM signal shows a maximum near the saturation intensity I_s:

$$\log(\%T) = \log(\%T)_o [1 + x(I/I_s)](1 + I/I_s) \qquad (40)$$

where $(\%T)_o$ is the transmission in the low power regime and x is unsaturated absorption loss.

Thus, regarding optical properties, it is evident that large photoconductivity, nonreflection of visible light and optical nonlinearity makes them promising materials for various optical media.

5.3 Magnetic Behavior of Composite Systems with Nanoparticles

The cooperative magnetic behavior of nanocomposites filled with magnetically ordered substance (e.g. Fe_2O_3) exhibits a number of anomalous effects caused by coherent rotation of all spins of such particles and thus the possibility of coherent change of direction for magnetization vector of the whole nanoparticle. In other words, the particle consists of only one domain. Two main patterns of behavior are possible in that situation: the absence of correlation between magnetization vectors of neighboring particles (which is named superparamagnetism), and the existence of such correlation (which is referred to as superferromagnetism).

α-Fe_2O_3 (hematite) nanocrystals of the size of ~ 20 nm, estimated by crypton adsorption both in the form of dispersed free particles and in silica matrix, were investigated using Mössbauer spectroscopy [10]. In these experiments the difference between superparamagnetic and superferromagnetic behavior was studied by comparing relaxation time for magnetization vector of nanoparticle and the time of Mössbauer measurement. If the relaxation time constant is greater than observation time, then the system exhibits superferromagnetic properties. It occurs when correlation bounds arise between the magnetization vectors of neighboring nanoparticles [71]. In the opposite situation, the magnetization vectors relax quicker than the mean observation time and this leads to averaging of the total magnetization vector and the absence of splitting of nucleus magnetic moment and, hence, to the absence of splitting in Mössbauer spectra.

According to superparamagnetic theory, a magnetic interaction may result in the ordering of magnetic moments of particles (these particles could be superparamagnetic under different conditions), the ordering temperature being [73, 74]:

$$T_p = \text{const} \frac{\mu_o M^2(T_p)}{k} \frac{d^6}{(d+S)^3}, \qquad (41)$$

where μ_o is the vacuum magnetic permeability, $M(T_p)$ is the particle magnetization vector at temperature T_p, d is the particle diameter and S is the distance between particles. Thus it is obvious that increasing interparticle distance S decreases the transition temperature, and the sample exhibiting superferromagnetic properties for free state is transformed into a superparamagnetic one when dispersed in polymer matrix at the same temperature.

Figure 23 shows the evolution of Mössbauer spectra with temperature for the free dispersed nanocrystals of α-Fe_2O_3 (A) and for the same nanocrystals in silica matrix (B). They exhibit clear superferromagnetic behavior for free nanoparticles, in clear sextet, which is characteristic of superferromagnetism, slightly disturbed by doublet on increasing the temperature from 77 to 300 K. In contrast, the composite of hematite with silica (B) shows "classic" superparamagnetic behavior reported in [75], in clear doublet at 300 K. In that case, with the temperature increasing more and more, particles transfer to the superparamagnetic state causing the doublet area to increase relative to the sextet one.

Thus, in accordance with Eq. (41), the characteristic temperature of magnetic transition differs for two samples by more than 200 K.

The question of the nature of correlation forces between neighboring particles is still open. In fact, two possibilities exist: magnetic dipole interaction

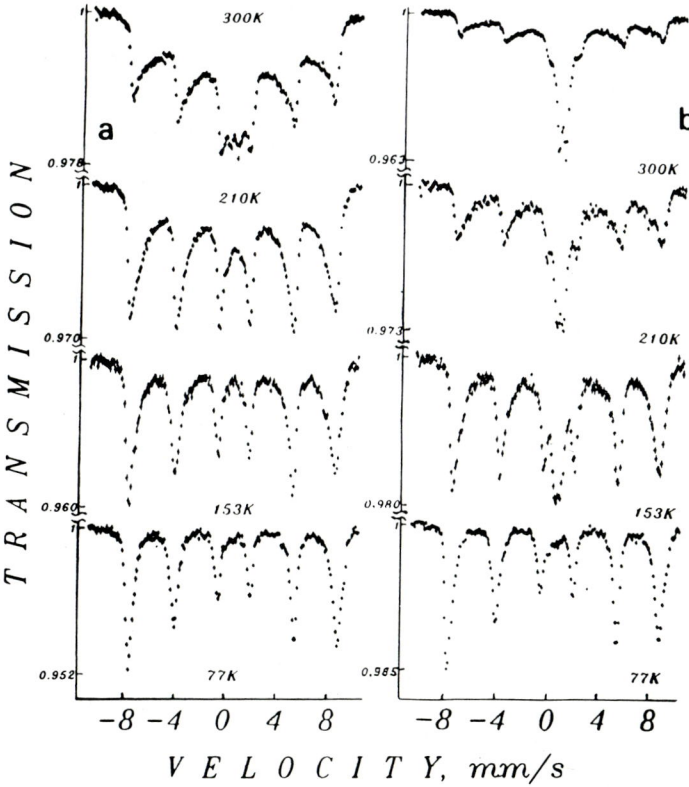

Fig. 24a. Mössbauer spectra of free dispersed hematite in superferromagnetic state in superparamagnetic state measured between T = 77 K and 300 K [10]. **b** Similar spectra for hematite in silica

and the exchange interaction between pairs of surface atoms belonging to two particles when they are in direct contact.

The other assumption which is of great interest from the theoretical point of view is whether there exists the concentration threshold, similar to the percolation threshold for electroconductivity of filled nanocomposites. In other words there is the range of magnetically ordered filler concentration in which the transition from superferromagnetic to the superparamagnetic state occurs at a certain temperature. Another question arising is which sort of excitations exist in such fractal-dimensional systems: what is the character of spin density wave in that case, and could there arise the nonlinear effects which are characteristic of one-dimensional systems. These questions are still open at present.

Conclusions

Summarizing the facts presented in this review one can stress some main points.

1. The polymer-nanocomposites possess some attractive features. In contrast to thin metal films and cermets they are three-dimensional objects containing nano-size particles. The polymer matrix prevents the oxidation of particles and coalescence, which is one of the reasons for instability and aging effects of free nanoparticle systems.

2. The combination of polymer matrix properties (such as electron transport and adsorption properties) with those of nano-particles make them promising materials for electrophotographic, photothermoplastic, sensor applications. They could also be useful as magnetics with high coercitive force.

3. The anomalous electromagnetic and optical properties arising in such systems near percolation threshold (at filler concentrations $\sim 10-25$ vol.%) connected with percolation threshold–the phase transition of the second type could be applied due to the nonlinear character of composite characteristics in the threshold range.

4. The composites with the concentrations both before and after percolation threshold combine the properties of polymer matrix (flexibility, processibility, durability etc.) with the properties of a semiconductor filler. These properties are not specific for composites with nanoparticles, but the small size of the filler provides for specific properties such as catalytic activity, which are not attainable for normal-size filler composites.

References

1. Wang Y, Mahler W (1987) Optical Communications 61: 233
2. Godovski DYu, Volkov AV, Karachevtsev IV Moskvino M. A., Volynskii A. L., and Bakeev N. F. (1993) Polymer Science USSR (Vysokomolekulyarnye Soedineniya) A 35: 1308

3. Parson T, Cupnick J, Davis M (1988) Phys. Rev. B, 38: 1282
4. Selby K, Vollmer M, Kresin V, deNeer WA, Knight WD (1989) Phys Rev B 40: 5417
5. Hass G, Francombe M, Hoffman R (eds) (1975) Physics of thin films, advances in research and development, vol. 8, Academic Press NY
6. Sheng P, Abeles B (1973), Phys Rev Lett. 31: 44.
7. Rosetti R, Ellison JL, Gibson JM and Brus LE (1984) J. Chem. Phys., 80: 4464–4469.
8. Henglein A, Kumar A, Janata J (1986) Chem. Phys. Lett. 132: 133
9. Arii T, Yatsuya A, Wada N, Mihama K (1978) Jap J Appl Phys, 17: 259–260.
10. Polikarpov M, Trushin I, Yakimov S (1992) J Magn Magn Mater, 116: 327–374
11. Morochov I, Trusov L, Lapovok V, Physical phenomena in ultradisperse media Energoatomizdat Publ. (1984) (in Russian).
12. Brus L. (1986) J Phys Chem. 90: 2555–2560
13. Wang Y and Herron N (1991) J. Phys. Chem. 95: 525–532
14. Ruschau GR, Newnham RE, Runt J, Smith BE (1989) Sensors and Actuators B 20: 269
15. Baral S, Fojtik A, Weller H, Henglein A, J Am Chem Soc 108: 375–378
16. Wang Y, Suna A, Mahler W et al. (1987) J Chem Phys, 87: 7315
17. Godovski DYu, Volkov AV, Moskvina MA et al (1994), the results are to be published in Polymer Science USSR A
18. Sandroff CJ. J. Phys. Chem., in print
19. Ekimov AI, Onushenko AA (1981) JETP Lett, 34: 346
20. Astor A, Staut E, J Phys Chem, in print
21. Nakamura N, Osashi S (1991) Inorg. Chem 30: 1788
22. Spanhel L and Anderson MA (1990) J Am Chem Soc 112: 2278
23. Fisher CH, and Henglein A (1989) J Phys Chem, 93: 5578
24. Fendler JH (1987) Chem. Rev. 87: 877
25. Xiao Kang and Fendler JH (1991) J Phys Chem. 95: 3716–3723
26. Stramel RD, Nakamura T and Tomas JK (1986) J Chem Phys Lett. 130: 423
27. Herron N, Wang Y, Eddy MM et al. (1989) J Am Chem Soc 111: 530 (1989)
28. Smotkin ES, Lee C, Bard AJ et al. (1989) Chem Phys Lett 152: 265
29. Dameron CT and Winge DR (1990) Inorg. Chem., 29: 1343
30. Dameron CT, Reese RN, Mehra RK et al. (1989) Nature 338: 596
31. Wang Y, Herron N (1988) Inorg Chem, 27: 435
32. Wang Y, Mahler W (1987) Opt Comm 61: 233
33. Stahanova SV, Nikonorova NI, Zanegin VD, Lukovkin GM, Volynski AL, Bakeev NF (1992) Polymer Science USSR 34: 133–140
34. Takahashi M, Miura H (1988) Top Curr Chem 144: 293
35. Akimov IA, Denisyuk IYu, Meshkov AM (1992) Russian Journal of Optics and Spectroscopy 72: 1026–1031
36. Hueberts G (1989) Top Curr Chem 145: 455
37. Tanaka T, Masashi M (1983) J Photog Sci 31: 13
38. Tanaka T, Ywasaki M (1985) J Imaging Sci 29: 86
39. Godovski DYu, Thesis, Moscow, 1993
40. Henglein A (1988) Top Curr Chem. 143: 113
41. Alivisatos AP, Harris A, Levinos N et al. (1988) J Chem Phys 89: 400
42. Dance IG, Choy A, Sendder ML (1984) J Am Chem Soc 106: 6285
43. Herron N, Wang Y, Eckert H (1990) J Am Chem Soc 112: 1322
44. Rafaeloff R, Tricot Y, Nome F (1985) J Phys Chem, 89: 533–537
45. Sejwick A, Sandos T (1990) J Phys Chem 92: 3454
46. Nickolas P, Roget R (1986) J Phys Chem 90: 271
47. Ozerin AA (1986) JETP Letters 39: 553
48. Marcus M, Flood W, Steigerwald M, Brus L. (1991) J Phys Chem 95: 1572–1576
49. Hu YZ, Koch SW, Lindberg M, Peygham Barian N, Pollock EL, Abiaham FF (1990) Phys. Rev. Lett. 64: 1805
50. Wang Y, Suna A, McHugh J et al. (1990) J Chem Phys, 92: 6927
51. Hanamyra E (1988) Phys Rev B, 38: 1228
52. Takagahara T (1987) Phys Rev B, 36: 9293
53. Migdal AB, Theory of finite Fermy-systems, GRFML Publishing, Moscow (1965) (in Russian).
54. Selby P. et al. (1991) Phys Rev B 43: 4565
55. Malov YuA, Zaretski DF (1993) Phys Rev Lett A 177: 379–383
56. Vollmer M and Kreibig U (1986) Proceedings of the 88th WE-Heraeus-Seminar: Nuclear

Concepts in the Study of Atomic Clusters Physics p. 266 Springer Verlag Publ. Berlin Heidelberg New York
57. Nedeljkovic JM, Nenadovic MT, Micic OI et al. (1986) J Phys Chem, 90: 12–13
58. Baral S, Fojtic A, Weller H and Henglein A, J Am Chem Soc 108: 375–378
59. Shklovski BI, Efros AL (1979) Physics of Doped Semiconductors, Nayka Publ., Moscow (in Russian).
60. Godovski DYu, Chmytin IA, Ponomarenko AA et al. (1994) Synthetic Metals 66: 19–23
61. Godovski DYu, Sukharev VYa, Volkov AV et al. (1993) Phys Chem Solids 54: 1613
62. Mott N, Twose R (1978) Conductivity of Noncrystalline Solids, Oxford Press.
63. Kirkpatrick S. (1961) Rev Mod Phys 10: 707
64. Godovki D, Sukharev V, Volkov A, Moskvina M, Sensors and Actuators B, in print.
65. Efros AL and Shklovskii BI (1976) Phys St Sol B, 76: 475–485.
66. Messinger J, vonRaben U, Chang R et al. (1981) Phys Rev B 24: 649
67. Qian SX, Snow JB, Chang RK in Laser Spectroscopy VII (1985) p. 204 Hanseh TW, Sher VR (ed) Springer-Verlag, Berlin Heidelberg New York.
68. Wang Y (1991) J Phys Chem 95: 1119–1124.
69. Leung KM (1986) Phys Rev A 33: 2461.
70. Godovski DYu, Volkov AV, Moskvina MA et al. (1994), Polymer Science USSR A, in print.
71. Yang W (1987) Opt Comm 61: 233
72. Yang W (1986) Opt Comm 60: 1657
73. Morup S, Madsen MB, Frank J, Villasden J, Koch CJW (1983) J Magn Magn Mater 40: 163
74. Morup S (1987) IEEE Trans Magn MAG-23: 3518.
75. Kundig W, Bommel H, Constabaris G and Lundquist RH (1966) Phys Rev 142: 327

Received: May 1994

Author Index Volumes 101-119

Author Index Vols. 1-100 see Vol. 100

Adolf, D. B. see Ediger, M. D..: Vol. 116, pp. 73-110.
Aharoni, S. M. and *Edwards, S. F.*: Rigid Polymer Networks. Vol. 118, pp. 1-231.
Améduri, B. and *Boutevin, B.*: Synthesis and Properties of Fluorinated Telechelic Monodispersed Compounds. Vol. 102, pp. 133-170.
Amselem, S. see Domb, A. J.: Vol. 107, pp. 93-142.
Arnold Jr., F. E. and *Arnold, F. E.*: Rigid-Rod Polymers and Molecular Composites. Vol. 117, pp. 257-296.
Arshady, R.: Polymer Synthesis via Activated Esters: A New Dimension of Creativity in Macromolecular Chemistry. Vol. 111, pp. 1-42.

Bahar, I., Erman, B. and *Monnerie, L.*: Effect of Molecular Structure on Local Chain Dynamics: Analytical Approaches and Computational Methods. Vol. 116, pp. 145-206.
Baltá-Calleja, F. J., González Arche, A., Ezquerra, T. A., Santa Cruz, C., Batallón, F., Frick, B. and *López Cabarcos, E.*: Structure and Properties of Ferroelectric Copolymers of Poly(vinylidene) Fluoride. Vol. 108, pp. 1-48.
Barshtein, G. R. and *Sabsai, O. Y.*: Compositions with Mineralorganic Fillers. Vol. 101, pp.1-28.
Batallán, F. see Baltá-Calleja, F. J.: Vol. 108, pp. 1-48.
Barton, J. see Hunkeler, D.: Vol. 112, pp. 115-134.
Berry, G. C.: Static and Dynamic Light Scattering on Moderately Concentraded Solutions: Isotropic Solutions of Flexible and Rodlike Chains and Nematic Solutions of Rodlike Chains. Vol. 114, pp. 233-290.
Bershtein, V. A. and *Ryzhov, V. A.*: Far Infrared Spectroscopy of Polymers. Vol. 114, pp. 43-122.
Bigg, D. M.: Thermal Conductivity of Heterophase Polymer Compositions. Vol. 119, pp. 1-30.
Binder, K.: Phase Transitions in Polymer Blends and Block Copolymer Melts: Some Recent Developments. Vol. 112, pp. 115-134.
Biswas, M. and *Mukherjee, A.*: Synthesis and Evaluation of Metal-Containing Polymers. Vol. 115, pp. 89-124.
Boutevin, B. and *Robin, J. J.*: Synthesis and Properties of Fluorinated Diols. Vol. 102. pp. 105-132.
Boutevin, B. see Amédouri, B.: Vol. 102, pp. 133-170.
Boyd, R. H.: Prediction of Polymer Crystal Structures and Properties. Vol. 116, pp. 1-26.
Bruza, K. J. see Kirchhoff, R. A.: Vol. 117, pp. 1-66.
Burban, J. H. see Cussler, E. L.: Vol. 110, pp. 67-80.

Candau, F. see Hunkeler, D.: Vol. 112, pp. 115-134.

Chen, P. see Jaffe, M.: Vol. 117, pp. 297-328.
Choe, E.-W. see Jaffe, M.: Vol. 117, pp. 297-328.
Chow, T. S.: Glassy State Relaxation and Deformation in Polymers. Vol. 103, pp. 149-190.
Chung, T.-S. see Jaffe, M.: Vol. 117, pp. 297-328.
Connell, J. W. see Hergenrother, P. M.: Vol. 117, pp. 67-110.
Curro, J.G. see Schweizer, K.S.: Vol. 116, pp. 319-378.
Cussler, E. L., Wang, K. L. and *Burban, J. H.*: Hydrogels as Separation Agents. Vol. 110, pp. 67-80.

Dimonie, M. V. see Hunkeler, D.: Vol. 112, pp. 115-134.
Dodd, L. R. and *Theodorou, D. N.*: Atomistic Monte Carlo Simulation and Continuum Mean Field Theory of the Structure and Equation of State Properties of Alkane and Polymer Melts. Vol. 116, pp. 249-282.
Doelker, E.: Cellulose Derivatives. Vol. 107, pp. 199-266.
Domb, A. J., Amselem, S., Shah, J. and *Maniar, M.*: Polyanhydrides: Synthesis and Characterization. Vol.107, pp. 93-142.
Dubrovskii, S. A. see Kazanskii, K. S.: Vol. 104, pp. 97-134.

Economy, J. and *Goranov, K.*: Thermotropic Liquid Crystalline Polymers for High Performance Applicaitons. Vol. 117, pp. 221-256.
Ediger M. D. and *Adolf, D. B.*: Brownian Dynamics Simulations of Local Polymer Dynamics. Vol. 116, pp. 73-110.
Edwards, S. F. see Aharoni, S. M.: Vol. 118, pp. 1-231.
Erman, B. see Bahar, I.: Vol. 116, pp. 145-206.
Ezquerra, T. A. see Baltá-Calleja, F. J.: Vol. 108, pp. 1-48.

Fendler, J.H.: Membrane-Mimetic Approach to Advanced Materials. Vol. 113, pp. 1-209.
Frick, B. see Baltá-Calleja, F. J.: Vol. 108, pp. 1-48.
Fridman, M. L.: see Terent'eva, J. P.: Vol. 101, pp. 29-64.

Geckeler, K. E. see Rivas, B.: Vol. 102, pp. 171-188.
Gehrke, S. H.: Synthesis, Equilibrium Swelling, Kinetics Permeability and Applications of Environmentally Responsive Gels. Vol. 110, pp. 81-144.
Godovsky, D. Y.: Electron Behavior and Magnetic Properties Polymer-Nanocomposites. Vol. 119, pp. 79-122.
González Arche, A. see Baltá-Calleja, F. J.: Vol. 108, pp. 1-48.
Goranov, K. see Economy, J.: Vol. 117, pp. 221-256.
Grosberg, A. and *Nechaev, S.*: Polymer Topology. Vol. 106, pp. 1-30.
Grubbs, R., Risse, W. and *Novac, B.*: The Development of Well-defined Catalysts for Ring-Opening Olefin Metathesis. Vol. 102, pp. 47-72.
van Gunsteren, W. F. see Gusev, A. A.: Vol. 116, pp. 207-248.
Gusev, A. A., Müller-Plathe, F., van Gunsteren, W. F. and *Suter, U. W.*: Dynamics of Small Molecules in Bulk Polymers. Vol. 116, pp. 207-248.
Guillot, J. see Hunkeler, D.: Vol. 112, pp. 115-134.
Guyot, A. and *Tauer, K.*: Reactive Surfactants in Emulsion Polymerization. Vol. 111, pp. 43-66.

Hall, H. K. see Penelle, J.: Vol. 102, pp. 73-104.
Hammouda, B.: SANS from Homogeneous Polymer Mixtures: A Unified Overview. Vol. 106, pp. 87-134.
Hedrick, J. L. see Hergenrother, P. M.: Vol. 117, pp. 67-110.
Heller, J.: Poly (Ortho Esters). Vol. 107, pp. 41-92.
Hemielec, A. A. see Hunkeler, D.: Vol. 112, pp. 115-134.
Hergenrother, P. M., Connell, J. W., Labadie, J. W. and *Hedrick, J. L.*: Poly(arylene ether)s Containing Heterocyclic Units. Vol. 117, pp. 67-110.
Hirasa, O. see Suzuki, M.: Vol. 110, pp. 241-262.
Hirotsu, S.: Coexistence of Phases and the Nature of First-Order Transition in Poly-N-isopropylacrylamide Gels. Vol. 110, pp. 1-26.
Hunkeler, D., Candau, F., Pichot, C.; Hemielec, A. E., Xie, T. Y., Barton, J., Vaskova, V., Guillot, J., Dimonie, M. V., Reichert, K. H.: Heterophase Polymerization: A Physical and Kinetic Comparision and Categorization. Vol. 112, pp. 115-134.

Ichikawa, T. see Yoshida, H.: Vol. 105, pp. 3-36.
Ilavsky, M.: Effect on Phase Transition on Swelling and Mechanical Behavior of Synthetic Hydrogels. Vol. 109, pp. 173-206.
Inomata, H. see Saito, S.: Vol. 106, pp. 207-232.
Irie, M.: Stimuli-Responsive Poly(N-isopropylacrylamide), Photo- and Chemical-Induced Phase Transitions. Vol. 110, pp. 49-66.
Ise, N. see Matsuoka, H.: Vol. 114, pp. 187-232.
Ivanov, A. E. see Zubov, V. P.: Vol. 104, pp. 135-176.

Jaffe, M., Chen, P., Choe, E.-W., Chung, T.-S. and *Makhija, S.*: High Performance Polymer Blends. Vol. 117, pp. 297-328.

Kaetsu, I.: Radiation Synthesis of Polymeric Materials for Biomedical and Biochemical Applications. Vol. 105, pp. 81-98.
Kammer, H. W., Kressler, H. and *Kummerloewe, C.*: Phase Behavior of Polymer Blends - Effects of Thermodynamics and Rheology. Vol. 106, pp. 31-86.
Kandyrin, L. B. and *Kuleznev, V. N.*: The Dependence of Viscosity on the Composition of Concentrated Dispersions and the Free Volume Concept of Disperse Systems. Vol. 103, pp. 103-148.
Kang, E. T., Neoh, K. G. and *Tan, K. L.*: X-Ray Photoelectron Spectroscopic Studies of Electroactive Polymers. Vol. 106, pp. 135-190.
Kazanskii, K. S. and *Dubrovskii, S. A.*: Chemistry and Physics of „Agricultural" Hydrogels. Vol. 104, pp. 97-134.
Kennedy, J. P. see Majoros, I.: Vol. 112, pp. 1-113.
Khokhlov, A., Starodybtzev, S. and *Vasilevskaya, V.*: Conformational Transitions of Polymer Gels: Theory and Experiment. Vol. 109, pp. 121-172.
Kilian, H. G. and *Pieper, T.*: Packing of Chain Segments. A Method for Describing X-Ray Patterns of Crystalline, Liquid Crystalline and none-Crystalline Polymers. Vol. 108, pp. 49-90.
Kokufuta, E.: Novel Applications for Stimulus-Sensitive Polymer Gels in the Preparation of Functional Immobilized Biocatalysts. Vol. 110, pp. 157-178.
Konno, M. see Saito, S.: Vol. 109, pp. 207-232.
Kressler, J. see Kammer, H. W.: Vol. 106, pp. 31-86.

Kirchhoff, R. A. and *Bruza, K. J.:* Polymers from Benzocyclobutenes. Vol. 117, pp. 1-66.
Kuleznev, V. N. see Kandyrin, L. B.: Vol. 103, pp. 103-148.
Kulichkhin, S. G. see Malkin, A. Y.: Vol. 101, pp. 217-258.
Kuchanov, S. I.: Modern Aspects of Quantitative Theory of Free-Radical Copolymerization. Vol. 103, pp. 1-102.
Kummerloewe, C. see Kammer, H. W.: Vol. 106, pp. 31-86.
Kuznetsova, N. P. see Samsonov, G. V.: Vol. 104, pp. 1-50.

Labadie, J. W. see Hergenrother, P. M.: Vol. 117, pp. 67-110.
Laso, M. see Leontidis, E.: Vol. 116, pp. 283-318.
Lazár, M. and Rychlý, R.: Oxidation of Hydrocarbon Polymers. Vol. 102, pp. 189-222.
Lenz, R. W.: Biodegradable Polymers. Vol. 107, pp. 1-40.
Leontidis, E., de Pablo, J. J., Laso, M. and *Suter, U. W.:* A Critical Evaluation of Novel Algorithms for the Off-Lattice Monte Carlo Simulation of Condensed Polymer Phases. Vol. 116, pp. 283-318.
Lesec, J. see Viovy, J.-L.: Vol. 114, pp. 1-42.
Liang, G. L. see Sumpter, B. G.: Vol. 116, pp. 27-72.
Lin, J. and *Sherrington, D. C.:* Recent Developments in the Synthesis, Thermostability and Liquid Crystal Properties of Aromatic Polyamides. Vol. 111, pp. 177-220.
López Cabarcos, E. see Baltá-Calleja, F. J.: Vol. 108, pp. 1-48.

Majoros, I., Nagy, A. and *Kennedy, J. P.:* Conventional and Living Carbocationic Polymerizations United. I. A Comprehensive Model and New Diagnostic Method to Probe the Machanism of Homopolymerizations. Vol. 112, pp. 1-113.
Makhija, S. see Jaffe, M.: Vol. 117, pp. 297-328.
Malkin, A. Y. and Kulichkhin, S. G.: Rheokinetics of Curing. Vol. 101, pp. 217-258.
Maniar, M. see Domb, A. J.: Vol. 107, pp. 93-142.
Matsuoka, H. and *Ise, N.:* Small-Angle and Ultra-Small Angle Scattering Study of the Ordered Structure in Polyelectrolyte Solutions and Colloidal Dispersions. Vol. 114, pp. 187-232.
Miyasaka, K.: PVA-Iodine Complexes: Formation, Structure and Properties. Vol. 108, pp. 91-130.
Monnerie, L. see Bahar, I.: Vol. 116, pp. 145-206.
Morishima, Y.: Photoinduced Electron Transfer in Amphiphilic Polyelectrolyte Systems. Vol. 104, pp. 51-96.
Müller-Plathe, F. see Gusev, A. A.: Vol. 116, pp. 207-248.
Mukerherjee, A. see Biswas, M.: Vol. 115, pp. 89-124.
Mylnikov, V.: Photoconducting Polymers. Vol. 115, pp. 1-88.

Nagy, A. see Majoros, I.: Vol. 112, pp. 1-113.
Nechaev, S. see Grosberg, A.: Vol. 106, pp. 1-30.
Neoh, K. G. see Kang, E. T.: Vol. 106, pp. 135-190.
Noid, D. W. see Sumpter, B. G.: Vol. 116, pp. 27-72.
Novac, B. see Grubbs, R.: Vol. 102, pp. 47-72.
Novikov, V. V. see Privalko, V. P.: Vol. 119, pp. 31-78.

Ogasawara, M.: Application of Pulse Radiolysis to the Study of Polymers and Polymerizations. Vol.105, pp.37-80.

Okada, M.: Ring-Opening Polymerization of Bicyclic and Spiro Compounds. Reactivities and Polymerization Mechanisms. Vol. 102, pp. 1-46.

Okano, T.: Molecular Design of Temperature-Responsive Polymers as Intelligent Materials. Vol. 110, pp. 179-198.

Onuki, A.: Theory of Phase Transition in Polymer Gels. Vol. 109, pp. 63-120.

Osad'ko, I.S.: Selective Spectroscopy of Chromophore Doped Polymers and Glasses. Vol. 114, pp. 123-186.

de Pablo, J. J. see Leontidis, E.: Vol. 116, pp. 283-318.

Padias, A. B. see Penelle, J.: Vol. 102, pp. 73-104.

Penelle, J., Hall, H. K., Padias, A. B. and *Tanaka, H.:* Captodative Olefins in Polymer Chemistry. Vol. 102, pp. 73-104.

Pieper, T. see Kilian, H. G.: Vol. 108, pp. 49-90.

Pichot, C. see Hunkeler, D.: Vol. 112, pp. 115-134.

Priddy, D. B.: Recent Advances in Styrene Polymerization. Vol. 111, pp. 67-114.

Privalko, V. P. and *Novikov, V. V.:* Model Treatments of the Heat Conductivity of Heterogeneous Polymers. Vol. 119, pp 31-78.

Pospisil, J.: Functionalized Oligomers and Polymers as Stabilizers for Conventional Polymers. Vol. 101, pp. 65-168.

Reichert, K. H. see Hunkeler, D.: Vol. 112, pp. 115-134.

Risse, W. see Grubbs, R.: Vol. 102, pp. 47-72.

Rivas, B. L. and *Geckeler, K. E.:* Synthesis and Metal Complexation of Poly(ethyleneimine) and Derivatives. Vol. 102, pp. 171-188.

Robin, J. J. see Boutevin, B.: Vol. 102, pp. 105-132.

Roe, R.-J.: MD Simulation Study of Glass Transition and Short Time Dynamics in Polymer Liquids. Vol. 116, pp. 111-114.

Rusanov, A. L.: Novel Bis (Naphtalic Anhydrides) and Their Polyheteroarylenes with Improved Processability. Vol. 111, pp. 115-176.

Rychlý, J. see Lazár, M.: Vol. 102, pp. 189-222.

Ryzhov, V.A. see Bershtein, V.A.: Vol. 114, pp. 43-122.

Sabsai, O. Y. see Barshtein, G. R.: Vol. 101, pp. 1-28.

Saburov, V. V. see Zubov, V. P.: Vol. 104, pp. 135-176.

Saito, S., Konno, M. and *Inomata, H.:* Volume Phase Transition of N-Alkylacrylamide Gels. Vol. 109, pp. 207-232.

Samsonov, G. V. and *Kuznetsova, N. P.:* Crosslinked Polyelectrolytes in Biology. Vol. 104, pp. 1-50.

Santa Cruz, C. see Baltá-Calleja, F. J.: Vol. 108, pp. 1-48.

Schweizer, K. S.: Prism Theory of the Structure, Thermodynamics, and Phase Transitions of Polymer Liquids and Alloys. Vol. 116, pp. 319-378.

Sefton, M. V. and *Stevenson, W. T. K.:* Microencapsulation of Live Animal Cells Using Polycrylates. Vol.107, pp. 143-198.

Shamanin, V. V.: Basesof the Axiomatic Theory of Addition Polymerization. Vol. 112, pp. 135-180.

Sherrington, D. C. see Lin, J.: Vol. 111, pp. 177-220.

Shibayama, M. see Tanaka, T.: Vol. 109, pp. 1-62.

Siegel, R. A.: Hydrophobic Weak Polyelectrolyte Gels: Studies of Swelling Equilibria and Kinetics. Vol. 109, pp. 233-268.
Singh, R. P. see *Sivaram, S.*: Vol. 101, pp. 169-216.
Sivaram, S. and *Singh, R. P.*: Degradation and Stabilization of Ethylene-Propylene Copolymers and Their Blends: A Critical Review. Vol. 101, pp. 169-216.
Stenzenberger, H. D.: Addition Polyimides. Vol. 117, pp. 165-220.
Stevenson, W. T. K. see *Sefton, M. V.*: Vol. 107, pp. 143-198.
Starodybtzev, S. see *Khokhlov, A.*: Vol. 109, pp. 121-172.
Sumpter, B. G., Noid, D. W., Liang, G. L. and *Wunderlich, B.*: Atomistic Dynamics of Macromolecular Crystals. Vol. 116, pp. 27-72.
Suter, U. W. see *Gusev, A. A.*: Vol. 116, pp. 207-248.
Suter, U. W. see *Leontidis, E.*: Vol. 116, pp. 283-318.
Suzuki, A.: Phase Transition in Gels of Sub-Millimeter Size Induced by Interaction with Stimuli. Vol. 110, pp. 199-240.
Suzuki, A. and *Hirasa, O.*: An Approach to Artifical Muscle by Polymer Gels due to Micro-Phase Separation. Vol. 110, pp. 241-262.

Tagawa, S.: Radiation Effects on Ion Beams on Polymers. Vol. 105, pp. 99-116.
Tan, K. L. see *Kang, E. T.*: Vol. 106, pp. 135-190.
Tanaka, T. see *Penelle, J.*: Vol. 102, pp. 73-104.
Tanaka, H. and *Shibayama, M.*: Phase Transition and Related Phenomena of Polymer Gels. Vol. 109, pp. 1-62.
Tauer, K. see *Guyot, A.*: Vol. 111, pp. 43-66.
Terent´eva, J. P. and *Fridman, M. L.*: Compositions Based on Aminoresins. Vol. 101, pp. 29-64.
Theodorou, D. N. see *Dodd, L. R.*: Vol. 116, pp. 249-282.
Tokita, M.: Friction Between Polymer Networks of Gels and Solvent. Vol. 110, pp. 27-48.

Vasilevskaya, V. see *Khokhlov, A.*, Vol. 109, pp. 121-172.
Vaskova, V. see *Hunkeler, D.*: Vol. 112, pp. 115-134.
Verdugo, P.: Polymer Gel Phase Transition in Condensation-Decondensation of Secretory Products. Vol. 110, pp. 145-156.
Viovy, J.-L. and *Lesec, J.*: Separation of Macromolecules in Gels: Permeation Chromatography and Electrophoresis. Vol. 114, pp. 1-42.
Volksen, W.: Condensation Polyimides: Synthesis, Solution Behavior, and Imidization Charcteristics. Vol. 117, pp. 111-164.

Wang, K. L. see *Cussler, E. L.*: Vol. 110, pp. 67-80.
Wunderlich, B. see *Sumpter, B. G.*: Vol. 116, pp. 27-72.

Xie, T. Y. see *Hunkeler, D.*: Vol. 112, pp. 115-134.

Yamaoka, H.: Polymer Materials for Fusion Reactors. Vol. 105, pp. 117-144.
Yoshida, H. and *Ichikawa, T.*: Electron Spin Studies of Free Radicals in Irradiated Polymers. Vol. 105, pp. 3-36.

Zubov, V. P., Ivanov, A. E. and *Saburov, V. V.*: Polymer-Coated Adsorbents for the Separation of Biopolymers and Particles. Vol. 104, pp. 135-176.

Subject Index

Absorption spectra 95
Auger microscopy 88

Bimetallic cluster catalysts 100
Boundary interphase 33, 52, 58-73
Bulk differential element 35
- representative element 34, 35, 49, 57, 60

Charge confinement 91
- transfer, interparticle 102
Cluster 81
-, infinite/isolated 52-57, 102
Cluster cataysts, bimetallic 100
Colloid solution, polymerization 83
Correlation length 54-56
Critical index 54-56

Degenerate-four-wave mixing 117
Differential cylindrical element 62

Electron-hole pair generation 103
Electroneutrality 90
Electrophotography 114
Equivalent element 58-61
EXAFS 87

Filled polymers 33, 66-71
Filler content, critical 58-60, 63-68, 73

Giant oscillator strength 98
Guarded hot plate device 23-24

Hashin-Shtrikman equation 44

Heat conductivity equations 37-73
- exchangers 27-28
- transport, unsteady-state 24-26
Herring equation 41
Hopping conductivity 105
-, mechanism 102
-, multiple 103

Inorganic filler particles 15, 16, 19, 20, 23
Internal network structure 6, 8, 12
Interparticle charge transfer 102
Ionization losses spectroscopy 88
- potential 99

Lichtenecker equation 41
Lorenz-Lorenz equation 37
Luminescence spectra 97

Malyshev-Malyshev equation 37
Maximum packing fraction 8-14, 16, 18, 20, 27
Maxwell equation 37
Meredith-Tobias equation 38
Metal particles 12, 14, 16, 18, 27
Metal-insulator threshold 102
Monte-Carlo simulation 107
Mossbauer measurement 118
Mott approach 107

Nanocomposites 81, 82
Nanoparticles, polylayer 85
Nanosize particle systems, synthesis 82-84

Optical absorption spectra 96

Particles, isolated   8-21
Percolation   52-57
- theory   102
Photosensitivity   114
Photothermoplasticity   114
Polymer blends   72, 73

Quantum size effects   89

Random resistor network   108
Rayleigh equation   38
- reflexion   88
Red shift, plasma frequency   99

Semiconductor chemical sensors   101
Shklovsky-Epfros approach   106
Simulation, Monte Carlo   107
Small angle X-ray analysis   85
Superferromagnetism   118
Superlattice   82
Superparamagnetism   118
Surface plasmon   99

Thermal conductivity   4-6, 23-24
Transmission electron microscopy   85

X-ray structural analysis   85

Springer-Verlag and the Environment

We at Springer-Verlag firmly believe that an international science publisher has a special obligation to the environment, and our corporate policies consistently reflect this conviction.

We also expect our business partners – paper mills, printers, packaging manufacturers, etc. – to commit themselves to using environmentally friendly materials and production processes.

The paper in this book is made from low- or no-chlorine pulp and is acid free, in conformance with international standards for paper permanency.

Printing: Saladruck, Berlin
Binding: Buchbinderei Lüderitz & Bauer, Berlin

42 2753

APR 6 '95